Amplifiers

SIMPLIFIED WITH 40 PROJECTS

No. 2885
$16.95

Amplifiers SIMPLIFIED WITH 40 PROJECTS

DELTON T. HORN

TAB BOOKS Inc.
Blue Ridge Summit, PA 17214

FIRST EDITION

FIRST PRINTING

Printed in the United States of America

Library of Congress Cataloging in Publication Data

Horn, Delton T.
Amplifiers simplified, with 40 projects.

Includes index.
1. Power amplifiers. I. Title.
TK7871.58.P6H67 1987 621.3815'35 87-10181
ISBN 0-8306-7885-9
ISBN 0-8306-2885-1 (pbk.)

Questions regarding the content of this book should be addressed to:

Reader Inquiry Branch
Editorial Department
TAB BOOKS Inc.
P.O. Box 40
Blue Ridge Summit, PA 17214

Contents

List of Projects

Introduction

COUNTLESS ELECTRONIC CIRCUITS EXIST FOR ALL SORTS OF applications. Despite the enormous variety of circuits, a few basic circuit types show up time and again.

Virtually every electronics system includes some form of amplifier circuit. An amplifier accepts a signal at its input and reproduces a replica of the original input signal at a larger amplitude at the output. This basic function is required in thousands of applications.

The amplifier is so important that it is mentioned in almost every text on electronics. However, specific information on a particular type of amplifier can often be hard to find. This book is intended to fill this need. It is a single-volume handbook on amplifiers, including both theory and practical considerations.

Forty projects and experiments are included to give you valuable hands-on experience. You can read this book straight through as a text or use it as a reference volume.

Chapter 1

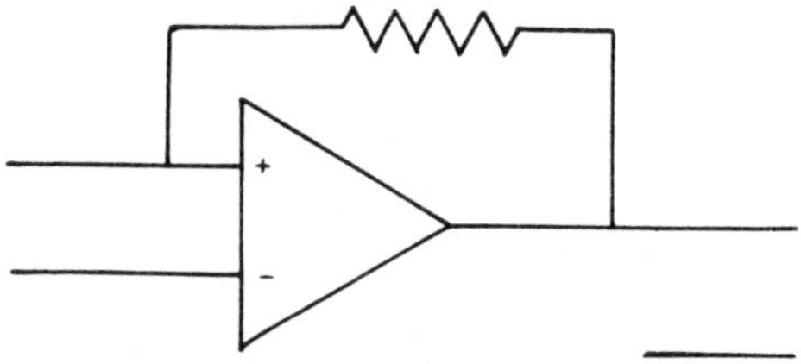

Amplifier Basics

MOST AMPLIFIER CIRCUITS INCREASE THE AMPLITUDE OF AN input signal, producing an output signal which is a larger replica of the input signal.

(Some amplifier circuits decrease the signal level, and should more properly be called attentuators; the principles of operation are the same. Other amplifier circuits have a gain of 1 or unity. The amplitude of the output signal is the same as the amplitude of the original input signal. These circuits are called buffer amplifiers and are used for isolation and impedance matching.)

This chapter explores the basic principles of amplifier design, primarily using bipolar transistors.

This book concentrates on solid-state (bipolar transistor, FET, IC, etc.) amplifiers. While some tube amplifiers are in use even today, they are by far the exception, rather than the rule.

THE COMMON-EMITTER CIRCUIT

Figure 1-1 shows the basic electrical connections for the correct biasing and normal operation of an npn transistor. While two voltage sources shown are here, most practical circuits don't actually have two independent power supplies; voltage sources are illustrated separately here for convenience in our discussion. Later, you'll learn how both of these biasing voltages can be obtained from a single power source.

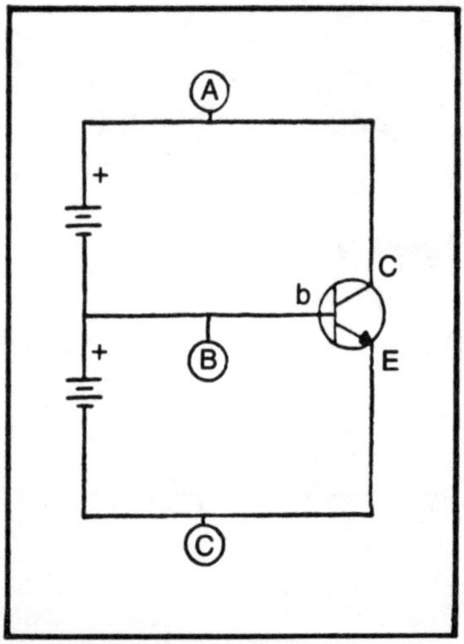

Fig. 1-1. The correct biasing for an npn transistor.

Take careful notice of the polarities within this circuit. The base is more positive than the emitter, but negative with respect to the collector. The actual voltage applied to the base can either be positive or negative with respect to the common ground point, but the polarity relationships between the transistor's various leads should always follow this pattern (unless you want the transistor to be cut off).

The negative charge from the terminal connected to the emitter forces the spare electrons in the N-type semiconductor of the emitter section of the transistor towards the base region. The base-emitter PN junction is forward biased. The extra electrons can cross over into the base section, filling the holes of the P-type semiconductor. There are more electrons moving into the base region than there are available holes. (The base section is considerably smaller than either the emitter or collector sections of transistor.) Because the base region has more electrons than in its normal state, it acquires an overall negative charge that forces the excess electrons out of this portion of the transistor.

A few of these electrons leave through the base lead to the positive terminal of the base-emitter battery, because the base lead is kept positive with respect to the emitter. The collector lead is even more positive, so most of the extra electrons are drawn out of the collector region, leaving it with a strong positive charge. This pulls most of the free electrons out of the base and into the collector, where they are drained off into the positive terminal of the base-collector battery.

All of this is summed up in the names given to the sections of a transistor—the emitter emits electrons and the collector collects them.

About 95 percent of the current flow passes through the collector, while only about 5 percent leaves the transistor via the base lead. The base current plus the collector current equals the emitter current. That is:

$$I_E = I_C + I_B$$

Just how much current is drawn by the emitter is determined by the characteristics of the specific transistor being used, and the amount of voltage being applied to the base terminal.

You can adapt the basic circuit of Fig. 1-1 to apply a variable voltage to the base, as shown in Fig. 1-2. The setting of the potentiometer (R1) determines the amount of voltage fed to the base, which, in turn, determines the amount of current drawn by the transistor. Regardless of the amount of current drain, only about 5 percent flows through the base lead, while the remaining 95 percent flows out the collector lead and through the load resistance (R2).

This theoretical circuit demonstrates that a very small change in the base current results in a very large change in the collector current. For this reason, transistors are often called current amplifiers. They amplify current, rather than voltage. Tubes, on the other hand, are voltage amplifiers.

Of course, thanks to Ohm's Law, the net effect is basically the same, since varying the current flow through the load resistance (R2), proportionately varies the voltage dropped across it.

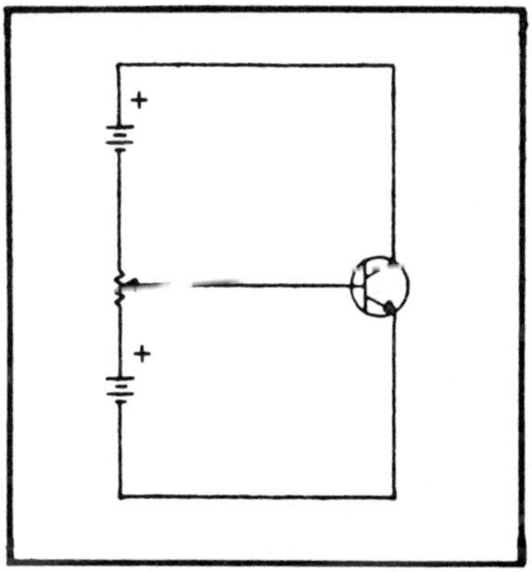

Fig. 1-2. Voltage applied to the base of a transistor controls collector current.

Figure 1-3 shows a more sophisticated version of this type of circuit. Notice that there is only one voltage source in this version. Resistors R1, R2, and R3 form a voltage divider to tap off the appropriate voltages at the correct relative polarities for proper biasing of the transistor.

If a very small ac voltage source is applied to the base, as shown here, the voltage on the base varies above and below its nominal dc value. This causes the collector current and thus the voltage dropped across the load resistance (R1) to vary in step with the ac voltage at the base. Because of the transistor's current gain, the ac voltage across R1 is significantly larger than the original ac voltage applied to the base. In other words, the ac signal at the input is amplified at the output. Because the emitter in Fig. 1-3 is at common ground potential, this type of amplifier circuit is called a common-emitter amplifier. The emitter is used as the common reference point for both the input and output signals.

In most practical common-emitter circuits, there is a resistor between the emitter and the actual common ground point. The resistor's function is to improve the stability of the circuit. The

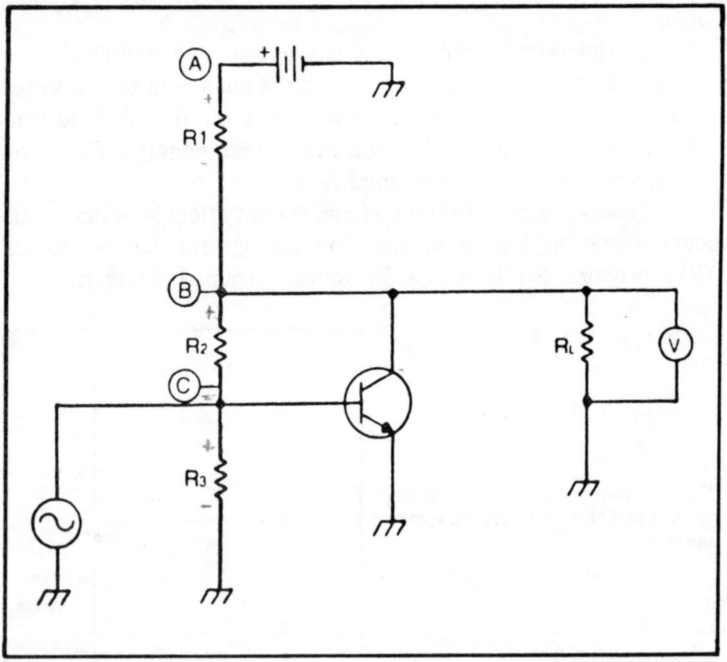

Fig. 1-3. The basic circuit of Fig. 1-1 can be adapted for use with a single-ended power supply.

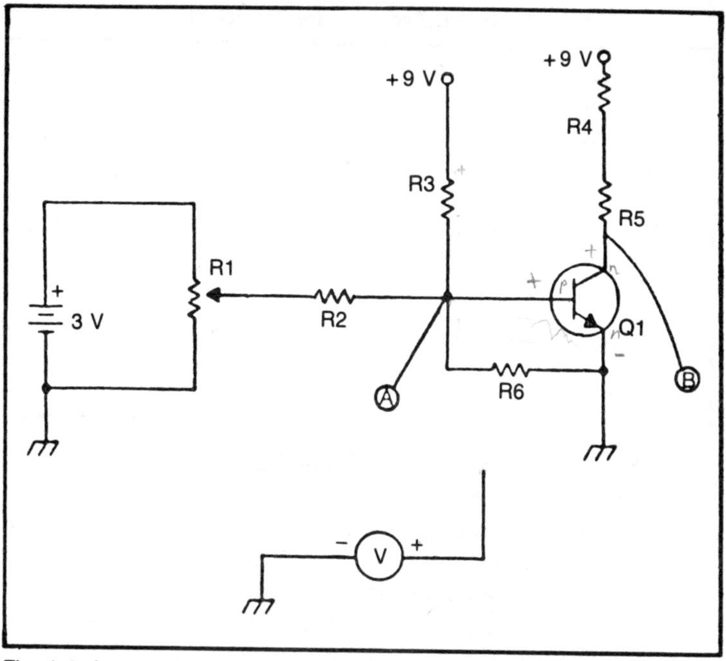

Fig. 1-4. An experimental common-emitter circuit (Project #1).

emitter is still considered grounded. The common-emitter amplifier configuration exhibits a low input impedance and a high output impedance. Current gain, voltage gain, and power gain are all respectably high. Common-emitter amplifiers always invert the signal. The output is always 180 degrees out of phase with the input. In other words, when the input signal goes positive, the output signal goes negative.

EXPERIMENTING WITH A COMMON-EMITTER AMPLIFIER CIRCUIT

A common-emitter circuit is illustrated in Fig. 1-4. Breadboard this circuit, and you can perform the following experimental tests yourself. A parts list is given in Table 1-1.

One lead of the voltmeter is left unconnected. If this lead is attached to point A, the meter displays the input voltage. If the lead is moved to point B, the output voltage will be indicated. Connect the free lead of the voltmeter to point A (input), and adjust the potentiometer until the input signal is exactly one volt.

Now move the voltmeter lead to point B and carefully measure the output voltage without touching the potentiometer. Enter this

Table 1-1. Parts List for Project #1.

Component Number	Description
Q1	npn transistor (2N3904 or similar)
R1	10 K potentiometer
R2	10 K resistor
R3	470 K resistor
R4, R5	1 K resistor
R6	100 K resistor
	solderless breadboarding socket
	voltmeter

value in the appropriate space in Table 1-2. Repeat this procedure for each of the input voltages listed in the table. When you are finished, examine the completed table carefully and notice how the output voltage varies in step with the input voltage.

The common-emitter amplifier is the most commonly used type of basic transistor circuit, but other configurations can be useful in certain applications.

THE COMMON-BASE CIRCUIT

Figure 1-5 shows a common-base amplifier circuit. Notice that the polarity relationships between the transistor leads remain the same as in the common-emitter configuration. The base is positive with respect to the emitter, but negative with respect to the collector.

The base is grounded, so its nominal value is zero volts. The collector is at a positive voltage (above common ground), and the emitter is at a negative voltage (below common ground). The voltage drop across resistor R2 causes the voltage applied to the emitter to be below ground potential (0 volts).

Resistor R3 and capacitor Cb are inserted between the actual base terminal and the ground point for stability. The values of these

Table 1-2. Worksheet for Project #1.

INPUT VOLTAGE	OUTPUT VOLTAGE
1 VOLT	____
1.5 VOLT	____
2 VOLT	____
2.5 VOLT	____

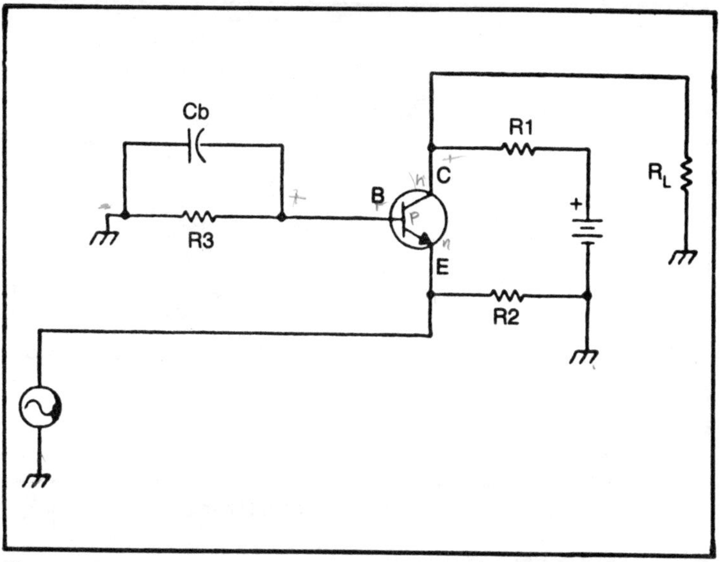

Fig. 1-5. A transistor can also be used in common-base configuration.

components are quite small to keep the voltage drop across them negligible. For all intents and purposes, the voltage applied to the base of the transistor is zero.

The power gain of a common-base amplifier is slightly higher than that of a common-emitter amplifier using the same transistor, but the voltage gain is significantly lower for the common-base circuit.

Another important difference between these circuit configurations is their input and output impedances. Remember that power is transferred between circuits most efficiently if the impedances match.

As already mentioned, the input impedance of a common-emitter amplifier is fairly low (typically between about 200 and 1000 ohms), and the output impedance is rather high (typically between about 10,000 to 100,000 ohms). The impedance of a common-base amplifier are similar, but the difference between the input and the output impedances are much more dramatic. The input impedance of a common-base amplifier is generally below 100 ohms, while the output impedance can be up to several hundred kilohms.

Another difference between these basic circuit configurations is that the output signal from a common-base amplifier is in phase with its input, while a common-emitter circuit inverts the input signal at the output (the signal is phase shifted 180 degrees).

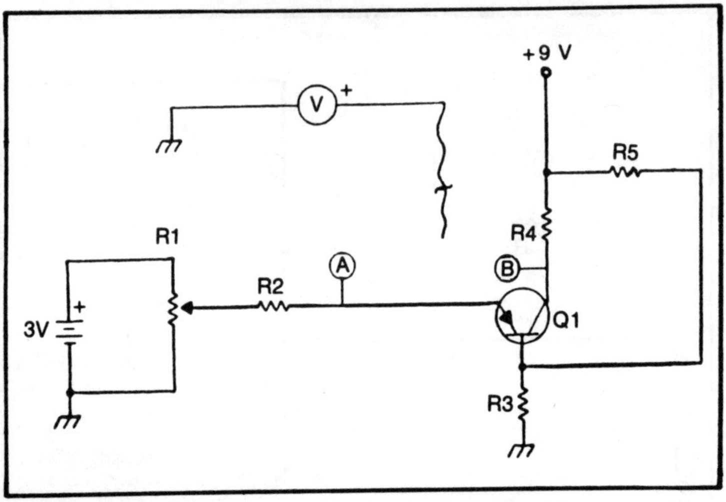

Fig. 1-6. An experimental common-base circuit (Project #2).

EXPERIMENTING WITH A COMMON-BASE AMPLIFIER CIRCUIT

Repeat the procedure of the last experiment with a common-base amplifier circuit shown in Fig. 1-6. The parts list for this experiment is given in Table 1-3. Enter your results for this experiment into Table 1-4.

THE COMMON-COLLECTOR CIRCUIT

As you probably have already guessed, the third transistor amplifier configuration uses a common-collector, as shown in Fig. 1-7. This circuit employs a positive ground point, so the operating voltages within the circuit are all negative.

Table 1-3. Parts List for Project #2.

Component Number	Description
Q1	pnp transistor (2N3906 or similar)
R1	10 K potentiometer
R2, R4	10 K resistor
R3	1 K resistor
R5	100 K resistor
	solderless breadboarding socket
	voltmeter

Table 1-4. Worksheet for Project #2.

INPUT VOLTAGE	OUTPUT VOLTAGES	
	NPN	PNP
1 VOLT 1 5VOLT 2 VOLT 2.5VOLT	____ ____ ____ ____	____ ____ ____ ____

The emitter is the most negative terminal. Resistor R1 drops some of this negative voltage so that the base is less negative (more positive) than the emitter, but it is still more negative than the collector. Thus the standard relative polarity requirements of the transistor are still met in this circuit.

One of the unique features of the common-collector amplifier configuration is that the voltage gain is always negative; that is, the output voltage is less than the input voltage. There is some positive power gain in this type of circuit, but it is relatively small compared to the power gains of the common-base and common-emitter configurations.

For obvious reasons, the common-collector circuit doesn't make a very good amplifier. It is quite a useful circuit for impedance

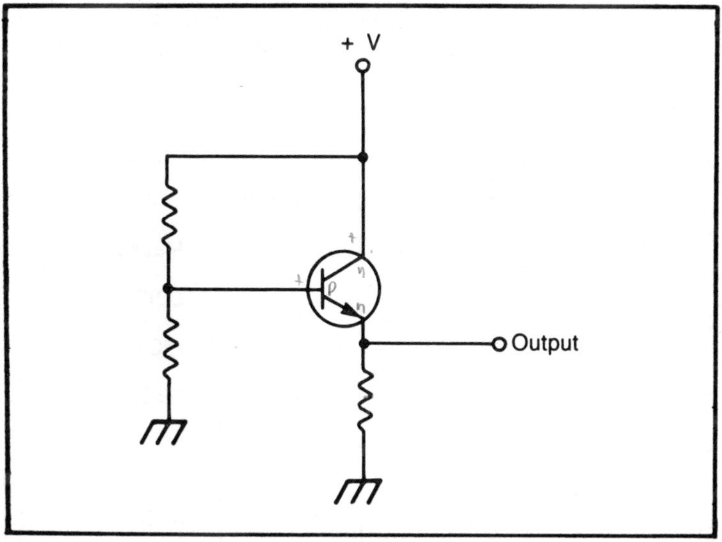

Fig. 1-7. The third basic transistor configuration is the common-collector circuit.

matching applications. With the other basic transistor configurations, the input impedance is always lower than the output impedance. In the common-collector circuit, the relative impedances are reversed. The output impedance is lower than the input impedance. The output of a common-collector amplifier is in phase with its input signal.

The difference between these three basic transistor circuit configurations are summarized in Table 1-5.

EXPERIMENTING WITH A COMMON-COLLECTOR AMPLIFIER CIRCUIT

The same experiment can be performed with the common-collector amplifier circuit of Fig. 1-8, using the parts listed in Table 1-6. Enter your results in Table 1-7.

Each of these experiments uses only npn transistors. Similar results are obtained with a comparable npn transistor, but all polarities are reversed.

AC AMPLIFIERS

Merely biasing a transistor at the quiescent operating point does not produce a functional circuit. The transistor must also be capable of amplifying an ac signal.

Direct current amplification is fairly straightforward, but if you need to amplify an ac (time-varying) signal, a number of additional

Table 1-5. Identifying Characteristics of the Three Basic Transistor Configurations.

COMMON ELEMENT	BASE	EMITTER	COLLECTOR
INPUT IMPEDANCE	VERY LOW	LOW	MEDIUM-HIGH
OUTPUT IMPEDANCE	VERY HIGH	HIGH	LOW
CURRENT GAIN	NEGATIVE	HIGH	HIGH
VOLTAGE GAIN	MEDIUM	HIGH	NEGATIVE
POWER GAIN	HIGH	HIGH	LOW
OUTPUT IN PHASE WITH INPUT?	YES	NO	YES

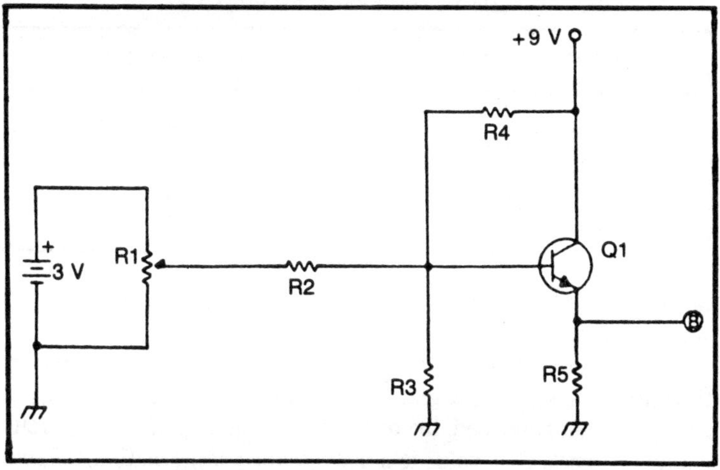

Fig. 1-8. An experimental common-collector circuit (Project #3).

factors must be considered to create a functional amplifier circuit. Some parameters, such as the gain or beta (β), may change from their nominal dc value when ac signals are used.

Circuit stability can become a problem, especially at higher frequencies. Several additional components must be added to a basic dc amplifier circuit to create a functional ac amplifier. These include a load resistor and several capacitors to prevent the ac signal from disturbing the dc bias voltages required for proper transistor operation.

Figure 1-9 shows a simple common-emitter ac amplifier circuit. Notice the overall similarity between this circuit and the common-emitter dc amplifier presented earlier. Besides the ac input signal source, just three components are added to the basic common-emitter circuit: capacitors C1 and C2, and load resistor R1. Resis-

Table 1-6. Parts List for Project #3.

Component Number	Description
Q1	npn transistor (2N3904 or similar)
R1	10 K potentiometer
R2, R5	1 K resistor
R3	100 ohm resistor
R4	470 K resistor
	solderless breadboarding socket
	voltmeter

Table 1-7. Worksheet for Project #3.

INPUT VOLTAGE	OUTPUT VOLTAGE
1 VOLT	____
1.5 VOLT	____
2 VOLT	____
2.5 VOLT	____

tance Rs represents the internal impedance of the input signal source itself.

Capacitor C1 blocks the base bias current provided by Rb from flowing back into the signal source and flowing through Rg to ground. If that happened, the quiescent operating point of the circuit would be upset. C1's value is normally fairly large so that ac input signals can pass through the capacitor to the base of the transistor for amplification.

The amplified output signal appears across collector resistor Rc, and is fed to the output and load resistor R1 through capacitor

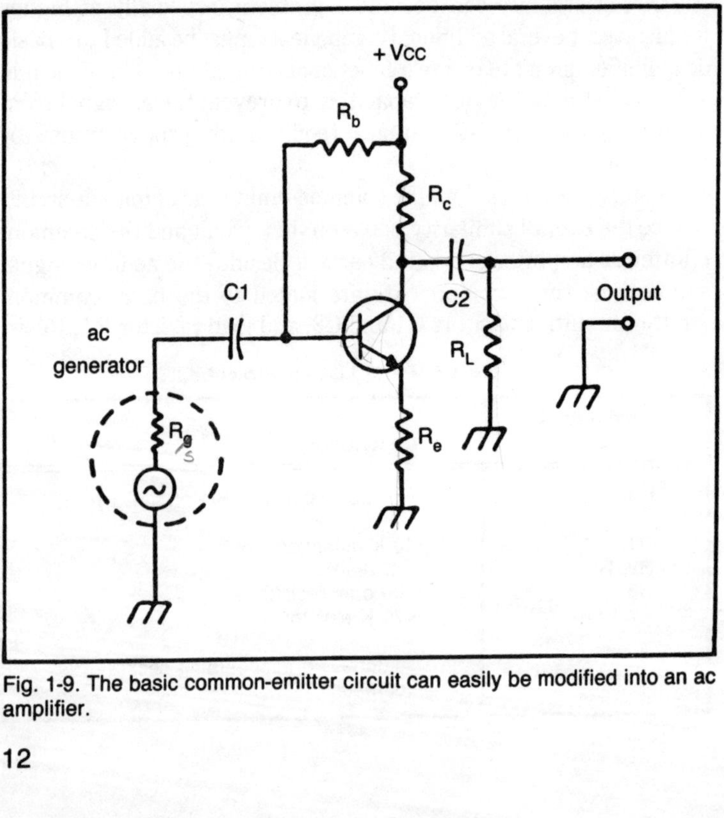

Fig. 1-9. The basic common-emitter circuit can easily be modified into an ac amplifier.

C2. This capacitor blocks the dc collector voltage from reaching the output. The amplified ac signal, however, passes through C2 to appear across load resistor R1.

This also works in the opposite direction. If capacitor C2 is not included in the circuit, the ac output signal can throw off the dc bias applied to the collector.

Resistor R1 represents the ac load on the output of the transistor. The output is the voltage across this resistor. R1 may not be an actual resistor in all circuits. It may represent the impedance of a loudspeaker, or the input impedance of an additional amplifier stage, or another circuit.

In a basic dc common-emitter amplifier, the output current is beta (β) times the input current. A similar relationship exists in the ac version, but here we must work with ac beta, which may or may not be equal to the dc beta value.

In technical literature, ac beta is sometimes referred to as h_{fe}. At other times, the Greek letter β is used to indicate the same quantity. To prevent unnecessary confusion, dc beta is commonly written as β_{dc}. This convention is used throughout this book. No subscript is used for ac beta, because it is a value normally used in design equations.

Dc beta can be found by taking the ratio of the collector and base currents:

$$\beta dc = \frac{I_C}{I_B}$$

The equation for ac beta is very similar:

$$\beta = \frac{\Delta I_c}{\Delta I_b}$$

The small triangles (read as "delta") indicate a changing (ac) value. In other words, ac beta is the ratio between the change in collector current (I_C) and the change in base current (I_B).

One practical way to find the value of ac beta is to use a set of characteristic curves, as shown in Fig. 1-10. The quiescent operating voltage (V_{ceq}) is selected, and the common point for the desired base current (I_{B3} in our example) and the appropriate

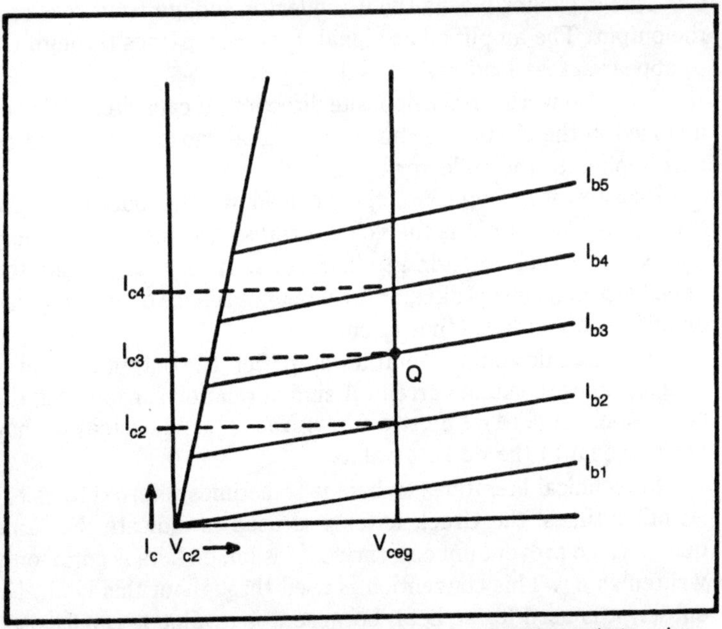

Fig. 1-10. A set of characteristic curves is used to find the ac beta value.

collector current (I_{C3}) in our example is found. This is marked as point Q in the diagram.

You can now find the value for ac beta by selecting additional crossing points on either side of Q, as illustrated in the sample diagram. The equation then becomes:

$$\beta = \left(\frac{I_{c4} - I_{c2}}{I_{b4} - I_{b2}} \right)$$

Similarly, ac alpha (α) is the ratio between the change in collector current and the change in emitter current:

$$\alpha = \frac{\Delta I_c}{\Delta I_e}$$

(From here on, dc alpha is written as α_{dc}. If just α is used, then ac alpha is assumed.)

It generally isn't very practical to try to determine ac alpha from the characteristic curves, because the value is , by definition, very close to unity (1). Fortunately, the relationship between ac alpha and ac beta is the same as for dc alpha and dc beta:

$$\alpha = \frac{\beta}{(1 + \beta)}$$

When working with an ac amplifier, you also have to consider some differences in the emitter resistance. For a dc circuit, this was just a discrete resistor (R_E) connected between the transistor's emitter and ground. This resistor is the only resistance of importance in the dc circuit. But when an ac signal is amplified by the transistor, the internal resistance of the emitter itself becomes significant. This internal resistance is identified in the equations as r_e. The small r indicates that it is an internal resistance, rather than a discrete resistor.

It is not too difficult to find the approximate value for the internal emitter resistance using this simple formula:

$$r_e = 26/I_C$$

where I_c is the collector current in milliamperes and r is the internal emitter resistance in ohms.

The total resistance in the emitter circuit is simply the sum of the internal emitter resistance and the value of the external emitter resistor:

$$R_{et} + R_e + r_e$$

As with the dc circuit, the emitter resistance (R_{et}) can be reflected back into the base circuit, where its value is multiplied by the ac beta.

The internal collector resistance is dependent on the circuit configuration used. In a common-base circuit, r_c is the internal collector-to-base resistance of the transistor. For common-emitter and common-collector circuits, the internal collector-to-base resistance is called r_d.

The easiest way to find the value of r_d is to use a set of characteristic curves, as shown in Fig. 1-11. Once the quiescent operating voltage (V_{ceq}) is determined, select two additional voltages that are on opposite sides of and equidistant to the

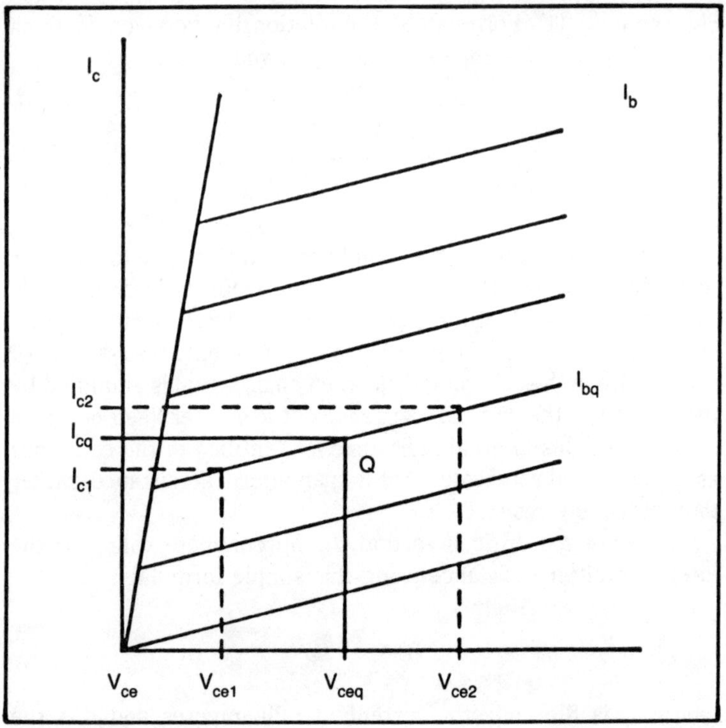

Fig. 1-11. A set of characteristic curves can also be used to find the ac collector resistance.

quiescent value. These additional voltages are labelled V_{cel} and V_{ce2} in the diagram. Find the collector current (I_{c1} and I_{c2}) for these two voltages, assuming that the base current (I_B) remains at its quiescent level. The value of the internal collector resistance (r_d) can now be found using these derived values:

$$r_d = \frac{V_{ce2} - VV_{cel}}{I_{c2} - I_{c1}}$$

A transistor also has an internal base resistance. However, this resistance is usually quite small (typically between about 500 and 1000 ohms), and can reasonably be ignored in the majority of design equations.

For the simple circuit shown in Fig. 1-9, the input impedance seen by the signal source works out to be:

$$Z_i = \frac{R_b[r_b + \beta(R_e + r_e)}{[R_b + r_b + \beta(R_e + r_e)]}$$

Since the value of r_b is generally negligable, you can simplify this equation to:

$$Z_i = \frac{[R_b \times \beta(R_e + r_e)]}{[R_b + \beta(R_e + r_e)]}$$

The output impedance seen looking back into the collector circuit is:

$$Z_o = \frac{r_d[(R_g + r_b) + \beta(R_e + re)]}{(R_g + r_b + R_e + r_e)}$$

The internal impedance of the signal source (rg) and the base resistance (r_b) are reflected from the input of the transistor to its output.

Once again, the r_b factor can usually be ignored, simplifying the equation:

$$Z_o = \frac{r_d[R_g + \beta(R_e + r_e)]}{(R_g + R_e + r_e)}$$

This formula actually gives the output impedance of the transistor itself. The output impedance seen by the load resistance (Rl) is actually this value in parallel with the collector resistor (Rc). That is:

$$R_{ol} = \frac{R_o \quad R_c}{R_o \; + \; R_c}$$

The current gain in the ac circuit (Ai) is the same as the ac beta value:

$$A_i = \beta$$

The formula for the ac voltage gain (Av) is somewhat more complex:

$$A_v = \left[\frac{R_l R_c}{(R_l + R_c} + \frac{1}{(R_e + r_e)} \right]$$

The ac power gain (G) is the product of the current gain and the voltage gain:

$$G = A_i A_v$$

FEEDBACK

Feedback is extremely important in establishing many circuit characteristics. In the simplest terms, feedback involves returning some of the output signal from a circuit back to the input of the circuit. Proper use of feedback can improve gain, bandwidth, and input and output impedances. On the other hand, feedback can cause circuit instability, clearly undesirable in an amplifier (although vital to an oscillator).

A signal that is fed back from the output of a circuit to its input so that the circuit's gain is reduced is called negative feedback. Conversely, a signal that is fed back so that the circuit's gain is increased, or so that the circuit breaks into oscillation, is called positive feedback.

In any circuit that uses feedback, such as the one shown in Fig. 1-12, a portion of (E_f) of the signal at the output (E_o) is fed back to the input. (The ratio E_f/E_o is often called B. Unfortunately, this can lead to confusion with the beta of a bipolar transistor, which is also identified by the symbol β. To try to minimize confusion, this book uses b when referring to the feedback ratio.) A voltage (Es) is applied to the input. With no feedback, the voltage at the input to amplifier is the signal from the source:

$$E_s = AE_{in}$$

The gain of the overall circuit is:

$$\frac{E_{out}}{E_{in}} = \frac{E_{out}}{E_s}$$

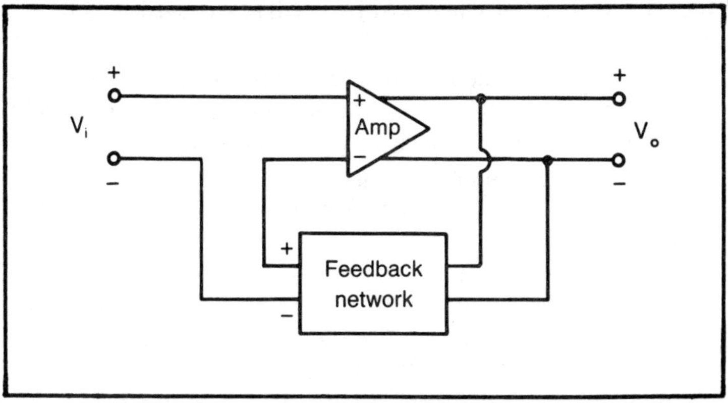

Fig. 1-12. Most practical amplifier circuits include a negative feedback loop.

But, when you add feedback to the circuit, some of Eout appears at the input as Ef. Assuming that the feedback is positive (in phase with the input signal), the input to the amplifier becomes:

$$E_{in} = E_s + E_f$$

When the feedback is negative, the signals are 180 degrees out of phase and the input to the amplifier is:

$$E_{in} = E_s - E_f$$

Since the level of Ef depends upon Eout and b, the gain of the circuit with feedback works out to:

$$A_f = \frac{A_v}{(1 - bA_v)}$$

where Av is the open-loop voltage gain of the amplifier without feedback, or:

$$A_v = \frac{E_{out}}{E_{in}}$$

Af is the gain of the overall circuit with feedback. The denominator of the equation for A_f $(1 - bA_v)$ is known as the feedback factor. When negative feedback is used, the expression

bA_v is negative, but when positive feedback is used, this term is positive.

Assuming that the feedback is negative and bA_v has a value much larger than 1, the gain of the amplifier is about equal to 1/b.

Negative feedback makes the gain of an amplifier less sensitive to variations in the supply voltage or other circuit parameters. On the other hand, positive feedback makes the gain more sensitive to such variations.

In this circuit, the feedback is applied in series with the original input signal, so the input impedance increases. The input impedance is increased by an amount that is directly proportional to the feedback factor. Conversely, the output impedance is reduced because the feedback signal is taken from across the load. The amount of this decrease is indirectly proportional to the feedback factor.

Negative feedback has other significant effects on the operation of the circuit. For example, it reduces distortion and improves frequency response of an audio amplifier. Distortion with feedback is equal to the distortion without feedback divided by the feedback factor.

In an amplifier with negative feedback, the high-frequency response is extended from Foh (the high frequency limit without feedback) to $F_{on}(1 - bA_v)$. Similarly, the low frequency response is extended from F_{ol} (without feedback) to $F_{ol}/(1 - bA_v)$.

Bear in mind here that since the feedback is negative, the bA_v terms are negative. Therefore:

$$1 - (-bA_v) = 1 + bA_v$$

The first step in designing an amplifier with feedback is to determine the amount of overall gain required. Assume you need a circuit with a voltage gain of 40 dB, or Av = 100. Also assume that in this application, approximately 20 dB of negative feedback is needed to reduce the distortion to about 10 percent of what it would have been without feedback. In this case, the gain of the overall circuit should be 100 × 10 = 1000 to preserve the effective circuit gain of 100 once the feedback has been applied.

If a resistor-capacitor network is included in the feedback path, the frequency response characteristics of the circuit can be altered. The RC network behaves as a filter.

Even when only resistors are used in the feedback path, the feedback response is not uniform over the entire frequency band. The frequency response varies with the overall gain of the circuit, as well as with the capacitance, inductance, and resistance inherent in the different sections of the circuit. These do not have to be actual physical components. For example, there are several internal capacitances between the leads of a transistor. These capacitances are discussed in detail in Chapter 7.

DISTORTION

In an ideal amplifier, the waveshape of the output signal is an exact replica of the input signal. Only the amplitude is changed. However, electronic circuit is ever perfect. Any amplifier circuit alters the waveshape at the output. In very high quality amplifiers, the alteration may be virtually undetectable, but it still exists.

This alteration of the waveshape is called distortion. There are many different types of distortion, but the most important type is harmonic distortion. In harmonic distortion, the circuitry adds additional frequency components to the output signal that are some multiple of some portion of the original input signal.

The sine wave is the only pure waveform with just a single frequency component (the fundamental). All other waveforms include one or more additional frequency components, which are generally harmonics, or whole number multiples of the fundamental frequency. The fundamental frequency defines the repetition rate of the waveshape.

If the fundamental is 250 Hz, the harmonic series is as follows:

2nd Harmonic = 500 Hz

3rd Harmonic = 750 Hz

4th Harmonic = 1000 Hz

5th Harmonic = 1250 Hz

6th Harmonic = 1500 Hz

7th Harmonic = 1750 Hz

8th Harmonic = 2000 Hz

9th Harmonic = 2250 Hz

10th Harmonic = 2500 Hz

The harmonic series continues indefinitely. As a general rule of thumb, the lower harmonics tend to be stronger and more significant than higher harmonics.

Not every complex waveform contains all possible harmonics. For example, symmetrical waveforms, such as the triangle wave and the square wave, contain no even harmonics, only the fundamental and odd harmonics. Non-symmetrical waveforms, such as rectangle waves and sawtooth waves, may contain a combination of odd and even harmonics along with the fundamental.

Any amplifier circuit tends to add to the harmonic content of the signal passing through it. The output can contain spurious frequency components not present in the original input signal. This is known as harmonic distortion. How much harmonic distortion an amplifier produces is indicated in a percentage specification called THD (Total Harmonic Distortion). The lower the percentage THD, the more distortion free the amplifier is.

Negative feedback can significantly reduce the THD of an amplifier. The lower harmonics are especially reduced by negative feedback. This is convenient, since they tend to be the most objectionable. High order harmonics are generally very weak, and can pass unnoticed. Often, the upper harmonics may be outside the audible range, making them totally insignificant in an audio amplifier.

If two or more separate musical waveforms are simultaneously passed through an amplifier, they tend to interact. The result of this interaction is called Intermodulation Distortion, or IM.

Other types of distortion also exist, but they are usually less of a problem than harmonic distortion. Various types of distortion, and methods of dealing with them, are presented throughout this book.

THE SIGNAL-TO-NOISE RATIO

Another important specification for determining the quality of an amplifier is the Signal-to-Noise ratio, usually abbreviated as S/N ratio. Any practical amplifier circuit generates some random noise. The S/N ratio defines the relative amplitudes of the desired signal and the undesired noise.

If you stretch the definition of distortion slightly and define it as the presence of any signal in the output that is not present at the input, then noise is a form of distortion. Where harmonic and intermodulation distortion are made up of discrete identifiable frequencies, noise usually contains varying amounts of random frequencies. (There are some exceptions, such as 60 Hz, hum.) Obviously, for an audio amplifier, any noise that is outside the audible spectrum (below approximately 20 Hz or above about 20 kHz) is of no particular concern, since it is inaudable.

In the past, noise measurements were normally made with low-pass and high-pass filters having cutoff points that eliminated the effects of any inaudible noise frequencies. It was soon discovered that just measuring the total noise content wasn't informative enough. Two amplifiers with same S/N rating could sound very different in terms of noise, the result of different concentrations of noise frequencies and the use of different weighting networks in making the noise measurements.

A weighting network is necessary for meaningful S/N measurements, because humans do not hear all frequencies at the same audible level when all frequencies are reproduced at an equal measured intensity. Two signals with the same amplitude but different frequencies won't sound equally loud.

Human ears tend to be less sensitive to frequencies at either end of the audible spectrum. Extreme bass and very high tones don't sound as loud as mid-range tones at a comparable amplitude. This is especially true at low listening levels. Since even in a moderate hi-fi system, the noise is reproduced at very low levels, it's clear that the non-linear response of the human ear plays an important role in evaluating the annoyance factor of noise. Some types of noise are more objectionable than others. To measure this nonlinearity in the frequency response of the human ear, the Fletcher-Munson curves were devised.

To compensate for these effects, three standard weighting networks have been devised for noise measurements. The standardization was made by the American National Standards Institute (ANSI). The standardized weighting networks are called simply class A, B, and C weighting.

To measure the internal noise of an amplifier, a reference output level is first defined by applying some known input to the device. In many cases, the rated output of the amplifier being tested is used for this reference output level.

The next step is to remove the signal from the input. The input is now grounded, and one of the weighting networks is inserted between the output and an average-reading ac voltmeter, as illustrated in Fig. 1-13. If the network has an insertion loss, this should be taken into account when the noise reading is taken. The

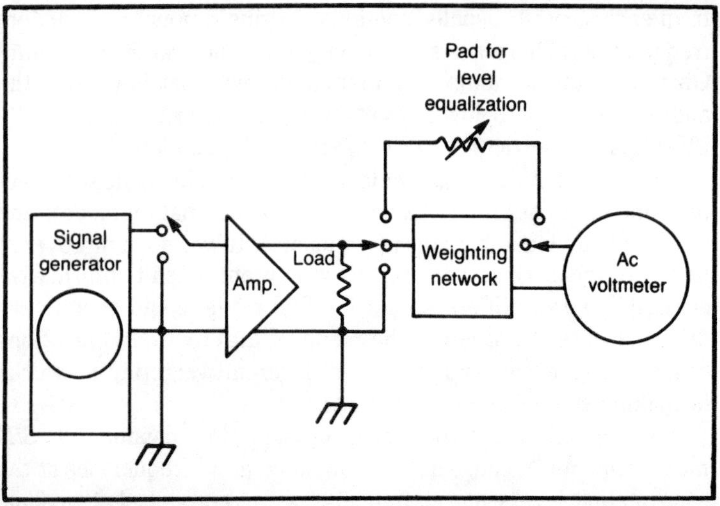

Fig. 1-13. To measure noise, a weighting network is placed between the amplifier's output and an ac voltmeter.

noise level is read from the voltmeter as "so many dB below the reference output." For a complete and truly meaningful specification, the type of weighting used should be identified.

The C weighing network has a frequency response with a continuous rolloff above 10 kHz and below 20 Hz. The rolloff rate is 6 dB per octave or more.

The B weighting network is quite similar to the C network, but the B network also has a simple high-pass network (filter) having its half-power (−3 dB) point at 160 Hz.

In the A weighting network, the frequency response of the C network is changed by the same amount as two simple identical nonisolated RC high-pass networks (filters) in cascade. Each of these has its half power (−3 dB) point at 280 Hz.

The frequency response of each of the three standardized weighting networks is shown in the plot of Fig. 1-14.

According to several psychoacoustic studies over the last few years, the so-called "annoyance factor" of noise does not necessarily correspond with the "equal loudness" curves of the older Fletcher-Munson studies. A more recent DIN network has been developed which differs somewhat from the standard A weighting network in that it includes a sharp cut-off filter above 9 kHz.

Additional changes in noise measurement have become necessary with the improved frequency response of modern hi-fi equipment. Older audio equipment generally cut off frequency

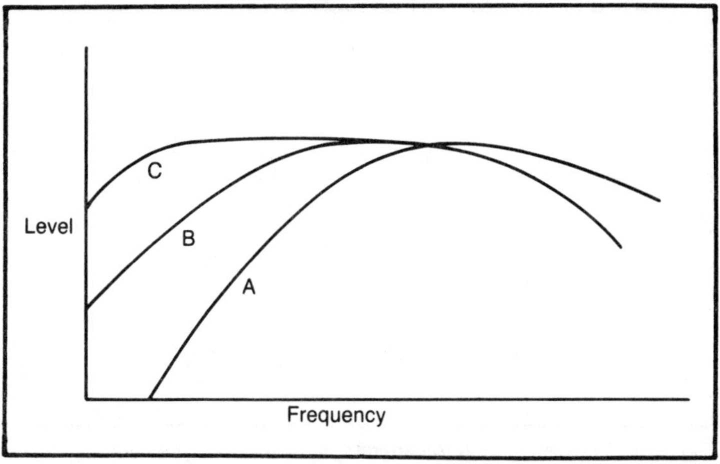

Fig. 1-14. A standard weighting network uses a non-flat frequency response to better correspond to the way the human ear hears different frequencies.

components above about 12 kHz to 15 kHz. The bandwidth of more modern audio devices has been extended to well beyond 15 kHz. High frequency noise is of more concern to today's audiophiles, so a "correct" weighting network (corresponding to annoyance factor) should take extreme high-frequency noise into account. The CCIR weighting network, shown in Fig. 1-15, is a popular choice for modern noise measurements.

TUBES VERSUS TRANSISTORS

Most modern electronic equipment is solid-state; that is, the active components are transistors or ICs. However, in certain

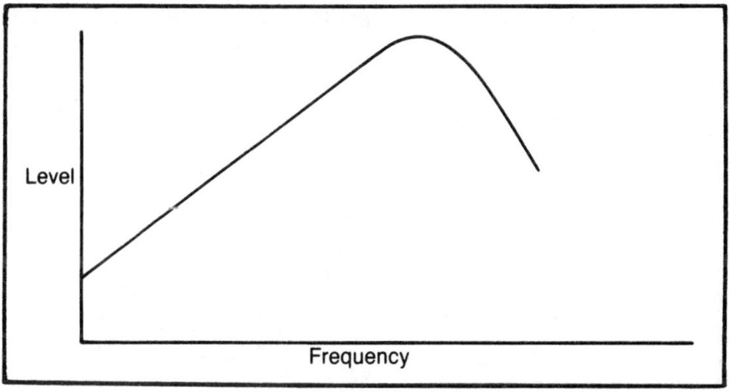

Fig. 1-15. One common form of weighting uses the CCIR network.

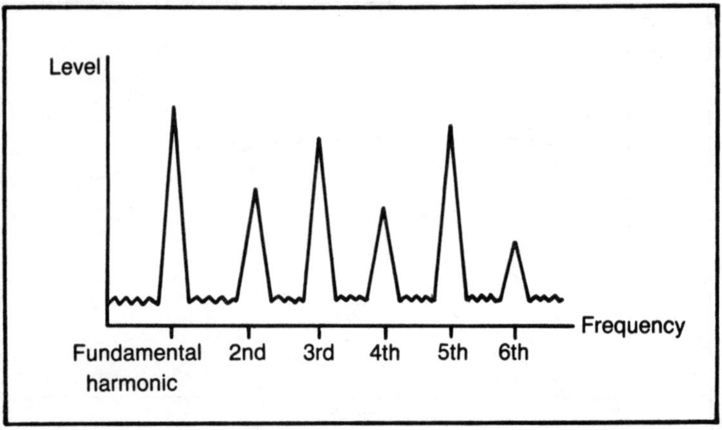

Fig. 1-16. Transistors tend to generate high-order odd harmonics.

audiophile circles, tube-based equipment has made a minor but significant comeback. Why is this?

Many audiophiles hear audible differences between tube and transistor amplifiers, even with similar specifications. The "tube sound" is generally considered warmer and more pleasing. This is primarily due to the manner in which bipolar transistors operate, which can lead to the generation of high-order odd harmonics. This can occur even at signal levels below clipping, as illustrated in Fig. 1-16.

These high-order harmonics tend to be more disturbing to listeners than lower-order harmonics. Of two amplifiers with the same THD rating, the one with stronger high-order harmonics will have a more objectionable sound. Whenever a signal overloads an amplifier, the way in which bipolar transistors exhibit clipping results in sharp, squared-off signals as shown in Fig. 1-17. Not

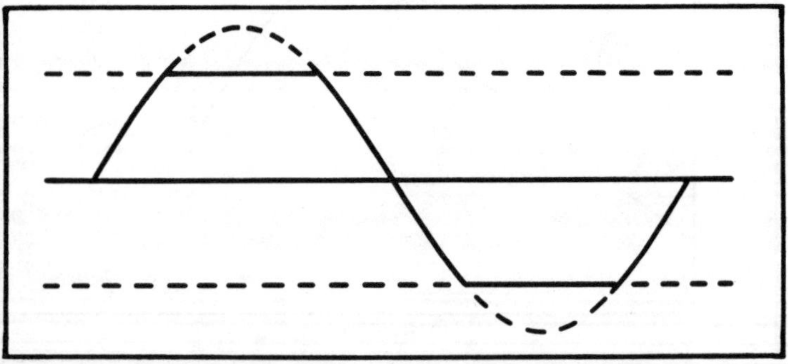

Fig. 1-17. Transistor clipping results in sharp squared-off signals.

surprisingly, such clipped signals contain large amounts of high-order harmonic distortion.

Tube and transistor amplifiers can also be distinguished in their transient response. Conventional bipolar transistors exhibit what is known as the carrier-storage effect. Imagine that the bipolar transistors in an amplifier are a series of storage tanks of unequal capacity, with a valve to shut off the flow. The output flow is regulated by the valve. Even if the valve is fully closed, isolating tank 1 from tank 2, output current continues to flow until tank 2 is fully drained into tank 3. This carrier-storage effect limits the rise and fall times of high-frequency transient signals. Even though the bandwidth of a circuit may extend far beyond the highest desired frequency, high frequency square waves may be distorted, as shown in Fig. 1-18.

The carrier-storage time is the interval between the beginning of the turn-off signal applied to the base of a saturated transistor and the instant that the collector voltage starts to rise toward the supply voltage. The carrier storage effect does not occur with tubes or FETs.

FETS

While bipolar transistors are the most common, there are a variety of other transistors. The FET, or Field Effect Transistor, is particularly important because its operation more closely resembles that of tube circuits than does the operation of bipolar transistors.

Transistors are current amplifiers, while FETs, like tubes, are voltage amplifiers. In other words, a bipolar transistor uses the input current to control the output current, while a FET uses a change

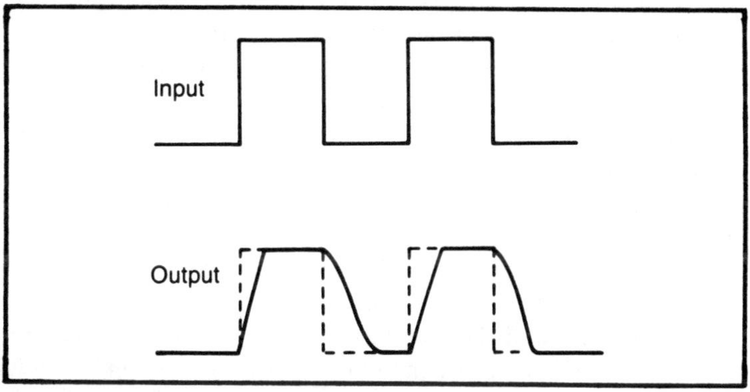

Fig. 1-18. The carrier storage effect alters a square wave.

in the input voltage to control the output current. Using an input voltage rather than an input current has a number of advantages that make the FET well-suited for use in audio amplifiers if it can produce a sufficiently high output current.

The carrier-storage effect exhibited by bipolar transistors is not present with FETs. Practical circuits using FETs are not really all that different from those using standard bipolar transistors. A FET has a much higher input impedance than a bipolar transistor. This means the FET does not contribute substantially to the load on a preceding stage, but the various capacitances within the FET itself can effect the load. The load for a FET amplifier stage can change the resonant frequency, when it is a tank circuit. Internal capacitances and their effects are most significant at higher operating frequencies. This subject is addressed more fully in connection with rf amplifiers in Chapter 7.

A bipolar transistor exhibits a direct relationship between temperature and current flow. If the temperature is raised, so is the amount of current flowing through the transistor. The more current flowing through any device, the hotter that device will become. This relationship can lead to a serious problem known as thermal runaway.

Suppose that for some reason, the transistor starts to draw a little too much current. This causes it to run hotter, forcing it to draw still more current, heating it up even more. This feedback-like cycle can continue until the transistor literally destroys itself. Special protection circuitry must be added to any power amplifier designed around bipolar transistors to avoid thermal runaway problems.

FETs, on the other hand, are immune to thermal runaway problems. The FET exhibits an inverse relationship between temperature and current flow. As the temperature increases, the FET draws less, rather than more, current. Consequently, a FET amplifier is essentially self-protecting.

Bipolar transistors are linear only up to a point. Any increase in voltage beyond a specific level produces no further increase in the output current. This condition is called saturation. FETs show no saturation in high-current operation. As the input voltage increases, the output current increases. This is quite similar to what happens in a triode tube circuit.

In light of all of these advantages, why aren't FETs used in all amplifiers? The unfortunate fact is that most power amplifier applications are just too much for the average FET to cope with.

Ordinary FETs just can't handle high currents. When the regulating input voltage is high the output current is limited and the device exhibits cut-off. This means that, as a rule, standard FETs are generally limited to voltage amplifying applications (such as rf amplifiers and low-level audio preamplifier stages) where the current flow is in the milliampere range.

Electronics is a field of constant change, and this situation has been altered somewhat in recent years with the development of newer, improved FET-like devices. In 1971, one solution to the FET power problem was created by Professor Nishizawa of Tohoku University. He obtained high output currents from a FET by drastically changing its internal structure. Nishizawa's invention was called the VFET, or Vertical FET.

In a VFET, the passage of current is still controlled by the input voltage applied to the gate, but the shape of the constriction is changed to permit higher current output, even with a high voltage input. The gate in a VFET has a comb-like structure that permits the current to pass through an infinite number of paths instead of the single path. To get an idea of the differences between a standard FET and a VFET, compare the construction of a standard FET (Fig. 1-19) with the construction of a VFET (Fig. 1-20).

In a standard, small-signal FET the current flows laterally or horizontally. In a VFET, however, current can flow vertically from source to drain. A vertical current path has far greater current

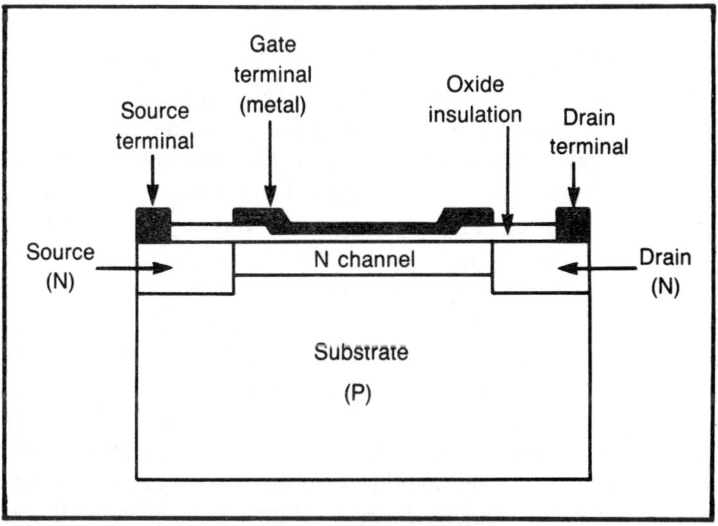

Fig. 1-19. In a standard FET the current flows laterally.

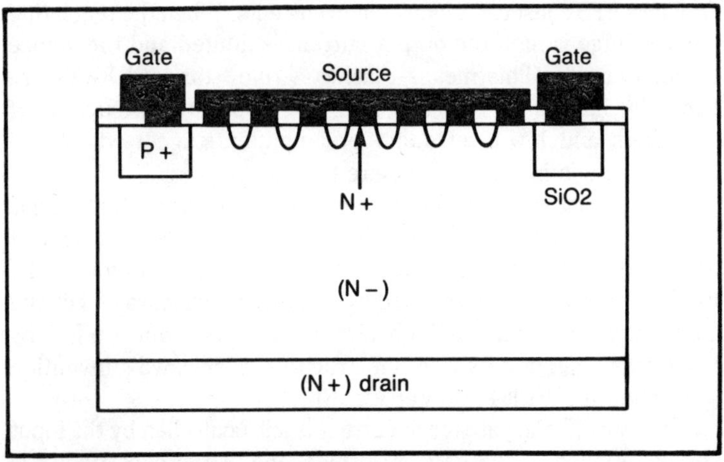

Fig. 1-20. In a V-FET the current flows vertically.

capacity and provides for a lower output resistance than a comparable horizontal current path. The current is controlled by the input voltage applied to the grid-like P+ gate(s).

The VFET does have a minor disadvantage in terms of size. The chip of a typical VFET is 3 mm square, while a typical small-signal FET chip measures only about 0.7 mm square.

A VFET's operation is very linear, which means that only small amounts of negative feedback are needed for wideband frequency response and a higher degree of stability.

TWO SIMPLE BUT PRACTICAL AMPLIFIER CIRCUITS

Here are a couple of practical general purpose amplifier circuits that can be used in many non-critical and low-power applications. The first of these circuits (Fig. 1-21), is a simple single transistor amplifier. This circuit can be used to boost almost any signal in the microvolt range. This is a very simple project that consists of so few components that it can actually be constructed to take up less space on a PC board than an IC.

This circuit features a gain of about 100 with the component values indicated in the diagram. Its frequency response is reasonably flat across the full audio spectrum.

Nothing in the design of this circuit is particularly critical. A wide range of substitutions in the component values can be made without significantly altering the circuit's performance. The capacitor values are particularly flexible. They may be changed if you don't happen to have the indicated values on hand. If your

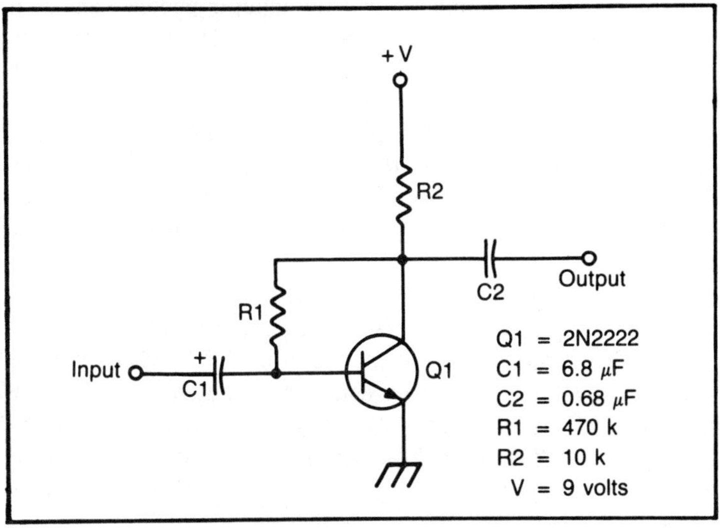

Fig. 1-21. Project #4 is a simple single-transistor amplifier.

application calls for a lower gain, simply reduce the value of feedback resistor, R1. Transistor Q1 can be almost any small signal high gain npn transistor. But remember, the selected transistor will have a major effect on the circuit gain. For example, a 2N2222 transistor can be used but the circuit will then have a lower gain than if the 2N3391 shown is used.

The second practical amplifier circuit is shown in Fig. 1-22. This circuit is rather unusual in that digital gates are employed for analog signal amplification. As you can see in the diagram, this amplifier circuit is comprised of three sections of a CMOS hex inverter (CD4049, or similar).

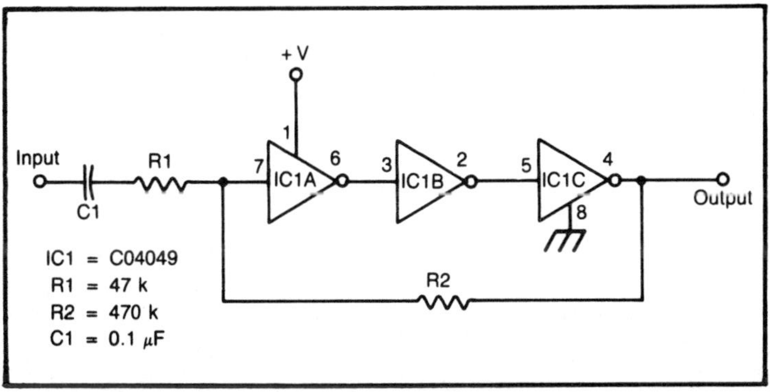

Fig. 1-22. Digital gates can be used for analog amplification (Project #5).

This amplifier exhibits a very high input impedance and so it does not load down the signal source. Since CMOS devices are used here, this circuit can operate on a wide range of power-supply voltages, with high noise immunity.

The gain of this circuit is determined entirely by the ratio of the feedback resistor value (R_2) to the input resistor value (R1). The frequency response is primarily a function of the input capacitor. The output signal swings from ground to a level just below the supply rail.

Unlike most linear amplifier circuits, this circuit's output voltage won't drop to ground potential when the input signal is removed. The output voltage always returns to V/2 (where V is the supply voltage) because we are using the digital inverters in a linear mode. This may or may not be a problem, depending on the specific application. If it does happen to be a problem in your particular application, you can correct it with a dc blocking capacitor at the output.

Chapter 2

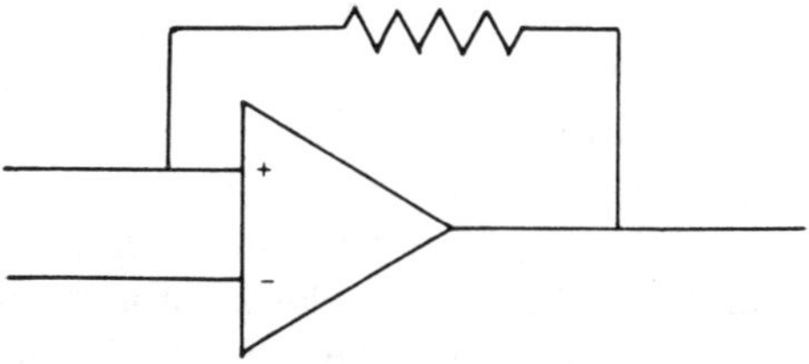

Amplifier Classes

AMPLIFIERS ARE DIVIDED INTO VARIOUS CLASSES. THIS CHAPter covers the most important of these amplifier classes and their differences. The discussion is confined to transistor amplifiers, but most of the information presented here also applies to comparable tube amplifiers.

Basically, in an amplifier circuit a transistor delivers power to the load. In the case of a dc amplifier, the power delivered to the load is simply equal to the load resistance multiplied by the square of the dc current flowing through that load:

$$P_L = R_L I_L^2$$

But things are more complex in an ac amplifier. If a full cycle appears across the load, the relationship of power to current and resistance remains the same as in a dc amplifier. But in an ac amplifier the effective current is the peak current in the cycle divided by the square root of 2, assuming that the ac signal in question is a sine wave. For other waveshapes the effective current is more difficult to calculate. This discussion assumes a sine wave whenever possible, just for the sake of convenience and clarity.

But what if there is less than a complete cycle passing through the amplifier? For example, if only one half of the cycle appears (with no current during the other half), the effective current is then equal to the peak current divided by two. If the signal is more than a half cycle, but less than a full cycle, the effective current falls somewhere between the two points mentioned above.

In an amplifier circuit, the portion of the cycle that appears across the output load is determined by the class of operation of the amplifier being used. Power amplifier classes range from Class A through Class H.

The most popular classes are A, B, AB, and C. The vast majority of audio and rf circuits are designed for operation in one of these classes. Classes D through H are used for special purposes, and special considerations are involved when designing a circuit to operate in one of these modes. For this reason, most of the attention is focused on Classes A through C.

Conceptually, the simplest kind of amplifier is a Class A type. In Class A circuits, the current flows through the transistor throughout the complete cycle. The transistor's bias is set at the center of the device's load line.

In Class B amplifiers, the current flows only during one half of a cycle and the transistor's bias current is set for $I_c = 0$. The transistor is cut off during the other half of the cycle.

Class AB is more or less a compromise mode. In this mode of operation, the current flows through the transistor for a little more than half of each cycle. It conducts less than in Class A, but more than in Class B.

Class C amplifiers conduct the least amount of current. In this mode of operation, current flows through the transistor for less than one half of each cycle.

Each of these amplifier classes is examined in more detail in the next few pages.

Most practical amplifiers are made up of multiple stages. Not all of the stages in a multi-stage amplifier are necessarily operating in the same class. Generally, when discussing power amplifiers, the design of the output stage is emphasized, since it is that stage that actually delivers power to the speakers. But several significant amplifier characteristics are determined by the earlier stages, particularly the driver or voltage-amplifier stages. These characteristics include frequency response, gain, thermal drift, and slew rate, among others. The driver stage may also have a great bearing on the distortion characteristics of the amplifier as a whole.

CLASS A AMPLIFIERS

In Class A amplifiers, the transistor is biased so that it conducts during the entire input cycle. This gives us an amplifier with very good linearity (low distortion), but low efficiency. Most of the power is consumed by the circuit. As much as 75 percent to 80 percent of the power input is wasted in this manner.

Generally, low power Class A amplifiers are biased at the center of their load line. Ideally, larger Class A power amplifiers are also biased at that point, but as the power increases, we have to consider some additional factors. It is an inescapable fact that for any transistor to deliver power, it must dissipate power. Despite the size of the load, the product of the collector current and collector-to-emitter voltage at any point on the load line must be less than the power dissipation rating of the transistor, to prevent it from self-destructing with the dissipated heat. To avoid such problems, the load line for the device should be selected so that it falls below the maximum permissible power dissipation curve of the unit. A simple Class A amplifier circuit is shown in Fig. 2-1.

Class A power amplifiers are widely used to feed power to a loudspeaker in small radios. Since most loudspeakers have a relatively low impedance—typically between 4 and 16 ohms—some sort of impedance matching network is usually employed to present a reasonable load to the collector circuit. In most cases, a transformer is used for this task.

Theoretically, when a transformer is used between the loudspeaker and transistor in a Class A amplifier, the maximum efficiency of the overall circuit is 50 percent. In other words, the maximum sinusoidal power that can be delivered to the load is equal

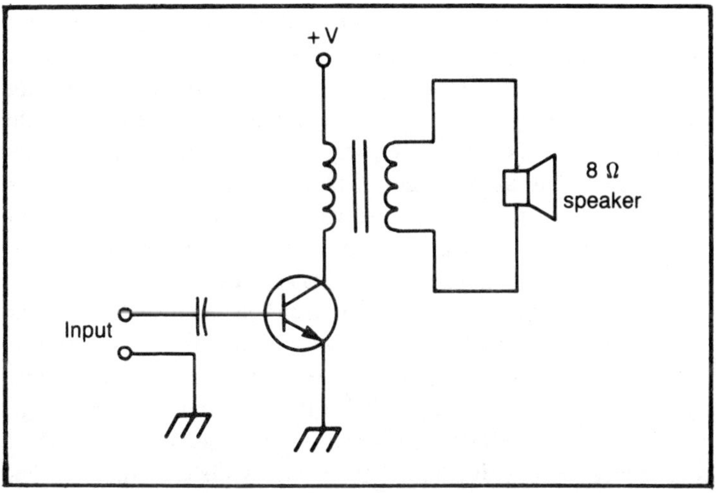

Fig. 2-1. A Class A amplifier conducts during the entire input cycle.

to 50 percent of the power that must be supplied by the source and dissipated by the transistor when idling. For example, if you bias the transistor so it demands a maximum of 10 watts from the power supply, only about 5 watts reaches the load over a cycle. The other 5 watts is dissipated by the transistor as heat.

An efficiency of only 50 percent certainly sounds very low, but the situation is usually even worse. In a practical Class A circuit, the efficiency is even 20 percent to 40 percent lower because of the losses and limits of the transformer and the circuit itself.

In some cases, matters can be worse yet. If the transformer is eliminated and the load is placed directly in the collector circuit, then the load has to dissipate dc power too. This limits the maximum efficiency to a mere 25 percent or so. And that 25 percent is still reduced by the same 20 percent to 40 percent mentioned earlier. The net result is an amplifier circuit with an efficiency of about 20 percent—with luck.

In short, the Class A amplifier circuit places a very high price on its excellent linearity. It wastes an incredible amount of power. Fortunately, there are more efficient ways to amplify ac signals.

CLASS B AMPLIFIERS

A Class B amplifier conducts for just half of each cycle, and rests the rest of the time. This makes for greater efficiency than the Class A amplifier is capable of. A simple Class B amplifier circuit is shown in Fig. 2-2.

To allow the Class B amplifier to conduct during just one half of a cycle, the transistor is biased at the point where the quiescent or idling current is equal to zero. If you assume that an npn transistor is being used, the only time current flows through the transistor is when the positive portion of the cycle is applied to the base. No current flows during the other half of the cycle, when the base is made negative. A pnp transistor works exactly the same way, of course, except that the polarities are reversed.

Unfortunately, chopping off half of each cycle is a very extreme form of distortion. This is clearly undesirable in virtually any audio application, especially if any degree of high fidelity is desired. In order for a Class B amplifier to reproduce signals with even reasonable accuracy, two parallel transistors must be used in the circuit. One of these parallel transistors is an npn type which reproduces the positive half of the cycle, while the other is a pnp type and reproduces the other half of the cycle. The two half cycles

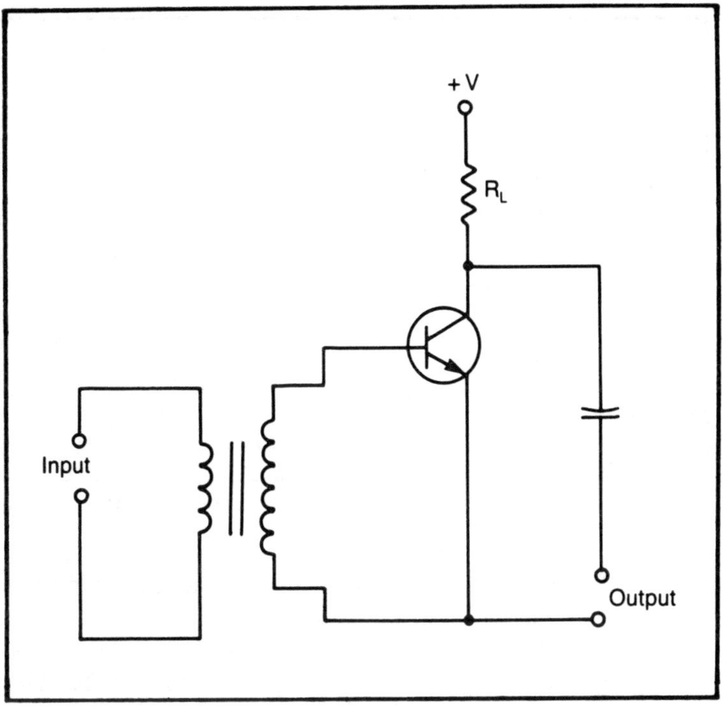

Fig. 2-2. A Class B amplifier conducts for only half of each input cycle.

are then recombined across the load to produce an amplified version of the original input waveform. This kind of circuit is called a push-pull amplifier. A typical push-pull circuit is illustrated in Fig. 2-3.

This circuit is far more efficient than the Class A amplifier discussed earlier. It can deliver up to 78.5 percent of the full cycle power demanded from the supply to the load. Because of the push-pull arrangement, the power delivered to the load over the complete cycle is double the maximum instantaneous power each individual transistor passes during the cycle. The amount of power a transistor actually dissipates depends on how much of the maximum output power is actually delivered to the load. When the power across the load is 40 percent of the maximum output signal that can be supplied by the circuit, maximum power is dissipated by the transistor. If the output is more or less than 40 percent of the maximum, the transistor demands less than the maximum power from the supply.

Since the two halves of each output cycle are being amplified by two separate transistors, any minute differences between the transistors (or their surrounding components) result in some degree of nonlinearity at the point where the two halves of the cycle

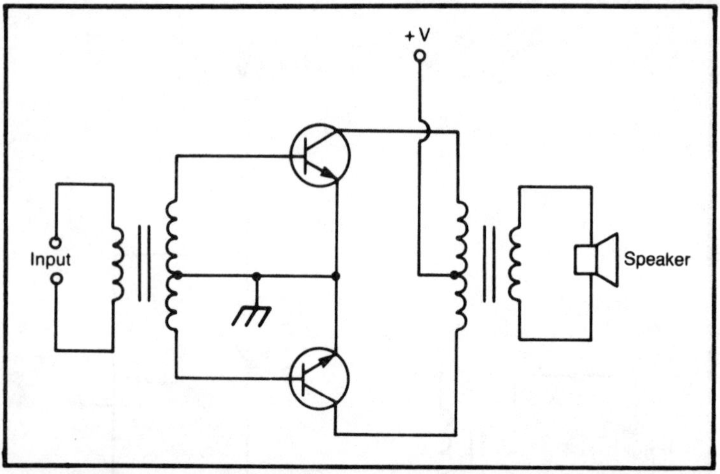

Fig. 2-3. To pass the entire waveform. Class B amplifiers are often connected in a push-pull configuration.

This is called crossover distortion. A signal with exaggerated crossover distortion is shown in Fig. 2-4.

Crossover distortion in Class B audio amplifiers can be reduced by adding a source of bias to the circuit, as shown in Fig. 2-5. This modification actually turns the circuit into a Class AB amplifier,

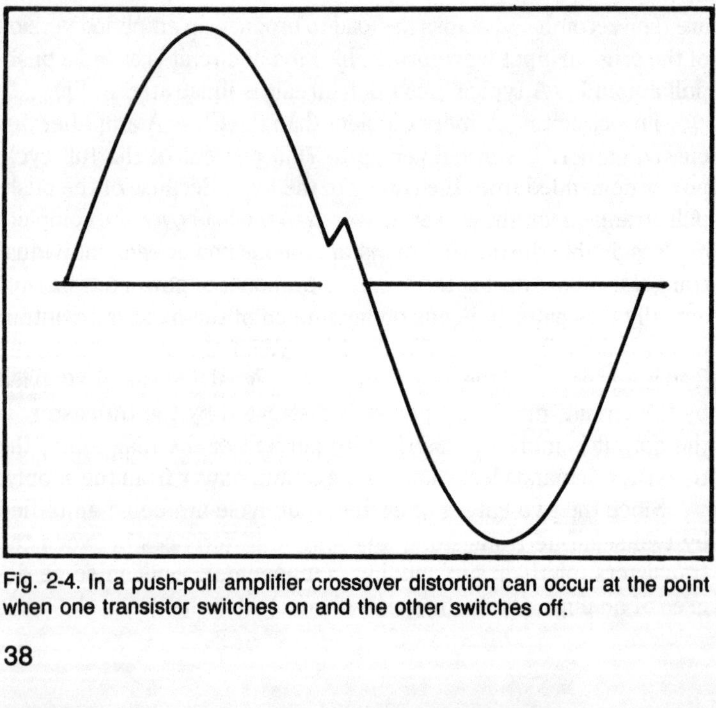

Fig. 2-4. In a push-pull amplifier crossover distortion can occur at the point when one transistor switches on and the other switches off.

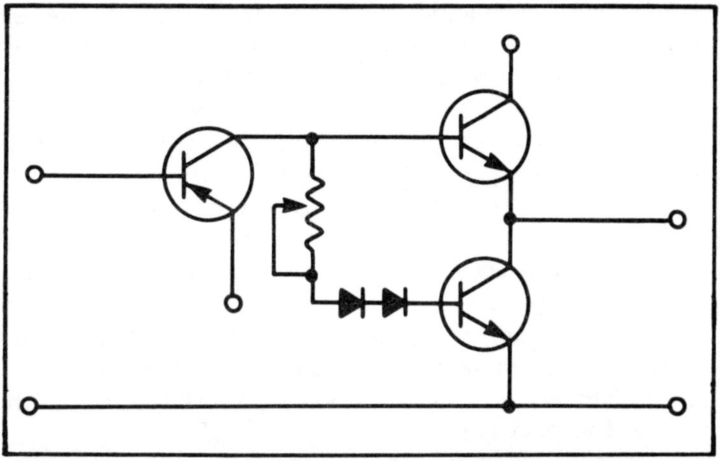

Fig. 2-5. Crossover distortion can be reduced with a suitable bias source.

since the collector current flows for more than half of each cycle. Class AB amplifiers are discussed in the next section of this chapter.

Because of its high efficiency, the Class B, push-pull direct coupled amplifier (as shown in Fig. 2-6) is probably the most popular

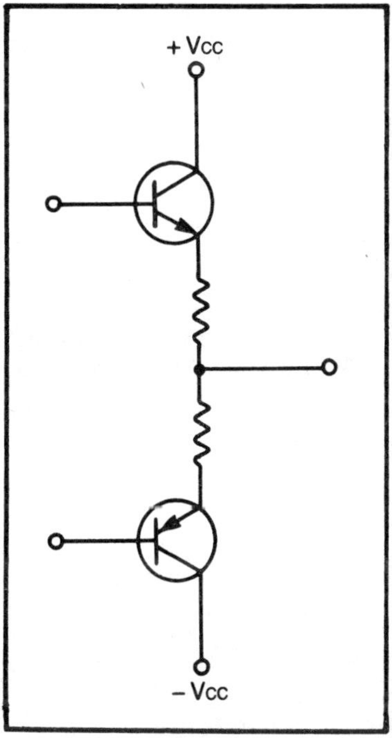

Fig. 2-6. Some Class B push-pull amplifiers use direct coupling.

audio-output configuration today. Early designs (before the mid-1970s) tended to rely more on capacitor coupling (illustrated in Fig. 2-7). The chief advantage of capacitor coupling is that such circuits require only a single ended power supply. In addition, since capacitors block dc voltages, speaker damage in the event of output failure was unlikely.

But designers and audiophiles soon learned that capacitor coupling tends to lead to poor low-frequency response and instability. Direct coupling is not susceptible to those particular problems, so it is now more popular. However, a direct coupled amplifier does require a dual polarity power supply, and complex circuits to protect the speakers against the possibility that one of the output transistors might short out and dump the full supply voltage across the speaker's voice coil.

With either direct or capacitor coupling, there are only two basic adjustments to be made:

☐ Quiescent (low or no signal) bias through the output transistors

☐ Output symmetry

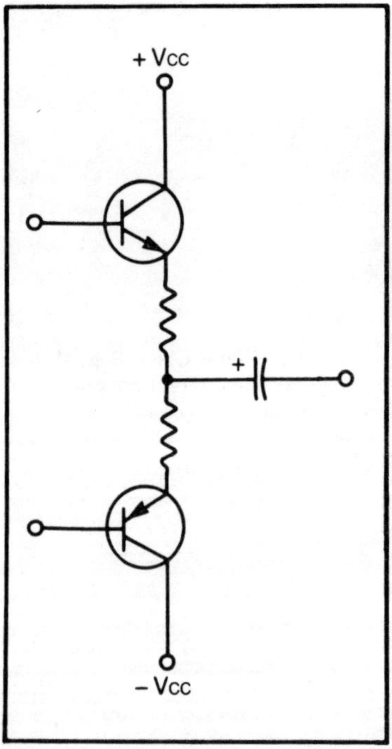

Fig. 2-7. Some Class B push-pull amplifiers use capacitor coupling.

The term "output symmetry" refers to how equally the positive and negative halves of the input waveform are reproduced. A misadjustment usually results in asymmetrical clipping and reduced output power for a given distortion rating.

While a Class B amplifier is highly efficient (up to nearly 80 percent efficiency) when it delivers its maximum power, it isn't quite as efficient at lower levels.

Under normal music listening conditions, an audio amplifier is generally driven to deliver full or nearly full output for only a very small fraction of the time it is operating. Under practical music listening conditions, the actual efficiency of a Class B amplifier can be as low as 20 percent.

CLASS AB AMPLIFIERS

The Class AB amplifier is a neat compromise between the linearity (low distortion) of a Class A amplifier and the efficiency of a Class B amplifier. Push-pull circuits are almost always used in Class AB. In Class A operation, the output transistor is biased at the center of the load line, while in Class B operation, the transistor is biased at the point where no collector current flows when the transistor is in the quiescent or idling state. In Class AB operation, the transistor is biased to operate somewhere between Class A and Class B.

In the AB mode, some small amount of collector current flows when the device is idling. That decreases the efficiency of the circuit somewhat, of course. But a Class AB amplifier can be useful where good and reliable reproduction of the input signal is required.

The chief advantage of the AB mode is that it significantly reduces crossover distortion. This type of distortion occurs in a Class B push-pull amplifier because no (or very little) conduction takes place through a transistor unless a specific minimum voltage is exceeded. Virtually no current flows through the base-emitter junction of a germanium transistor if less than 0.2 or 0.3 volt is applied. For a silicon transistor the minimum signal voltage is between 0.5 and 0.7 volts.

As the ac input signal passing through a Class B push amplifier goes from positive to negative (or vice versa) there is a brief delay between the time one transistor is cut off and the other is turned on. This slight delay is what causes crossover distortion. If a transistor is cut off at a rapid rate, large transient voltages develop in the circuit. This could cause the transistor to break down. Such problems are pretty much avoided in Class AB operation.

CLASS C AMPLIFIERS

In a Class C amplifier, the transistor is biased so that it conducts for less than one half of each cycle. Only a very small portion of each cycle passed through the amplifier. Of course, this severely distorts the signal at the output, but in some applications (especially rf amplifiers) linearity isn't nearly as important as low power consumption. Class C amplifier circuits are extremely efficient. The output of a Class C amplifier is in the form of pulses, and the efficiency can be as high as 85 percent.

Using an npn transistor, a Class C amplifier is created by biasing the base so that it is negative with respect to the emitter.

The Class C mode of operation is absolutely worthless for audio applications, but it is widely used in i-f and rf circuits. A resonant LC circuit is normally placed at the output. Each time a signal is applied to the amplifier's input, a full cycle is generated across the LC circuit, provided that the circuit is tuned for the frequency (or a multiple of the frequency) of the signal applied to the amplifier's input.

HIGHER CLASSES

In recent years, a number of manufacturers of high fidelity equipment have developed new approaches to amplification, trying to achieve the best balance between fidelity and efficiency. These newer amplifier circuits have been dubbed Classes D through H. Generally, the usefulness of these modes is limited mainly to specialized circuits and applications. In some cases, the original manufacturer holds a patent on the class. These new amplifier classes are briefly discussed in the following sections.

Class D Amplifiers

The Class D amplifier was developed by Infinity Systems, Inc. In the Class D mode, an amplifier employs high frequency pulses (greater than 200 kHz) that are first modulated by the audio signal to be amplified and then decoded by an integrating circuit that restores the audio envelope (waveshape).

When the audio signal is converted into a train of pulses, the width of each pulse is related to the instantaneous amplitude of the audio waveform. When reproduced through an amplifier, the pulses are fed to a push-pull circuit. The output from this circuit is connected through a resistor to a loudspeaker that is shunted by a capacitor, as shown in Fig. 2-8.

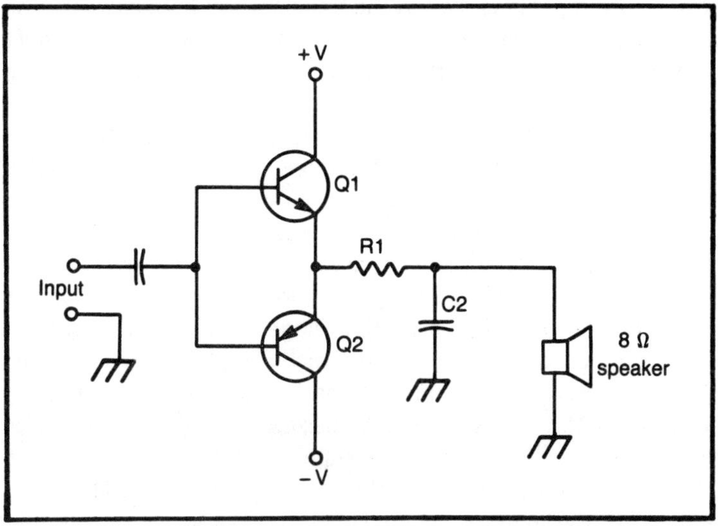

Fig. 2-8. A Class D amplifier.

Notice that no transformer is used between the load and transistors. In this type of circuit the transistors feed the amplified pulse signals directly to the loudspeaker through resistor R1. R1 and capacitor C1 form an integrator (filter) circuit where the width of each pulse determines the voltage developed across capacitor C2. The voltage across C2 appears across the loudspeaker. This converts the pulses back to audio waveforms, which are heard through the speaker.

Since the duty cycle of each high frequency pulse is very short, the output transistors never conduct for long at a time. Excess heat doesn't get a chance to build up. In Class D operation, heat dissipation is a fraction of that encountered with more conventional Class B circuits. Because there is relatively little energy being wasted as heat, the overall efficiency (at least when the amplifier delivers close to its maximum power output) is quite high.

Class D amplifiers are also called switching amplifiers.

Class E, F, and G Amplifiers

Classes E, F, and G amplifier circuits are very similar to one another. Each of these classes is based on the push-pull circuit. In each half of the push-pull circuit, two transistors are connected in series. When the input signal is low, only one of the transistors in each half of the push-pull circuit conducts current, while no voltage is applied to the remaining devices. However, when the signal level at the input gets higher, the transistors in both halves of the push-pull circuits are turned on. Since all the transistors conduct only during the peaks in the audio input signal, efficiency is quite high.

Let's discuss the Class G amplifier in a little more depth. Classes E and F work in pretty much the same way, but there are some differences that we won't bother to go into here.

The Class G was developed by the Hitachi Company of Japan. A basic Class G circuit is illustrated in Fig. 2-9. Class G was designed to enable amplifiers to operate more efficiently over a larger percentage of their total operating range. This is based upon the way amplifiers are called upon to actually amplify musical signals.

The signal to be amplified, marked V_{in} in the diagram, is applied to the bases of the transistors. Load resistor R_1 is connected to emitter of Q_1. The supply voltage (V_1) is applied to the collector of Q_1 and the emitter of Q_2, through diode D_1. At the same time, the collector of Q_2 is connected to a second supply voltage (VCC). VCC is higher than supply voltage V_1. Diode D_1 also functions as an isolation device. It prevents any current from flowing from VCC from flowing back into V_1.

If the input level (V_{in}), is less than the first supply voltage (V_1) then Q_2 is cut off because its base-emitter junction is reverse

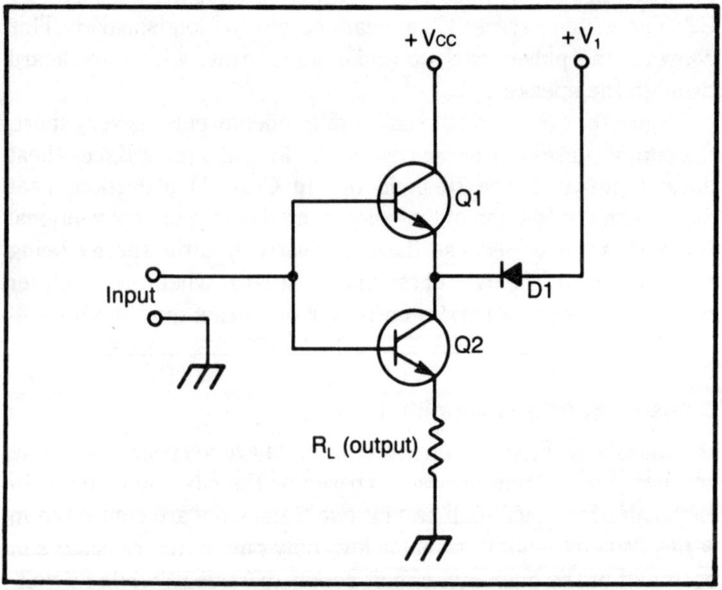

Fig. 2-9. The Class G amplifier is a fairly recent development.

biased. The current flowing through R_1 is supplied from V_1 through D_1. Under these conditions, instantaneous efficiency is equal to V_{in}/V_1.

Now let's consider what happens when the input signal (V_{in}) is at a higher level than the first supply voltage (V_1) (but less than VCC). Now transistor Q_2 becomes forward biased and is turned on. The current flowing through R_1 is supplied from VCC through Q_2.

Ignoring the saturation voltage between collector and emitter of Q_2 for convenience, the instantaneous efficiency is V_{in}/VCC.

The two efficiency levels of the Class G amplifier are shown in Fig. 2-10. The vertical line represents the point at which the supply voltage transition takes place.

A Class G amplifier is almost two amplifiers in one. For low-level input signals, the efficiency of this circuit is improved considerably and the amount of heat generated in the output transistor is reduced compared with conventional Class B amplifiers.

The Class G amplifier is not perfect. There are certain problems inherent in the basic form of this mode of operation. For one thing, a form of crossover distortion occurs at the point when transistor Q_2 is turned on or off. The circuit should be modified so that Q_1 is not saturated until Q_2 is switched on. This modification helps prevent this distortion. It is accomplished by adding diode D_2 in Fig. 2-11. D_2 may be a zener diode or it can even be a simple resistor since it only has to maintain a voltage equal to the V_{be} voltage of Q_2.

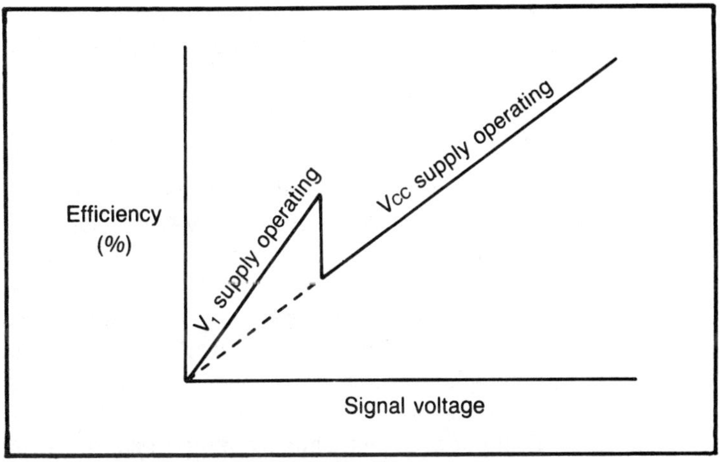

Fig. 2-10. A Class G amplifier has two efficiency levels.

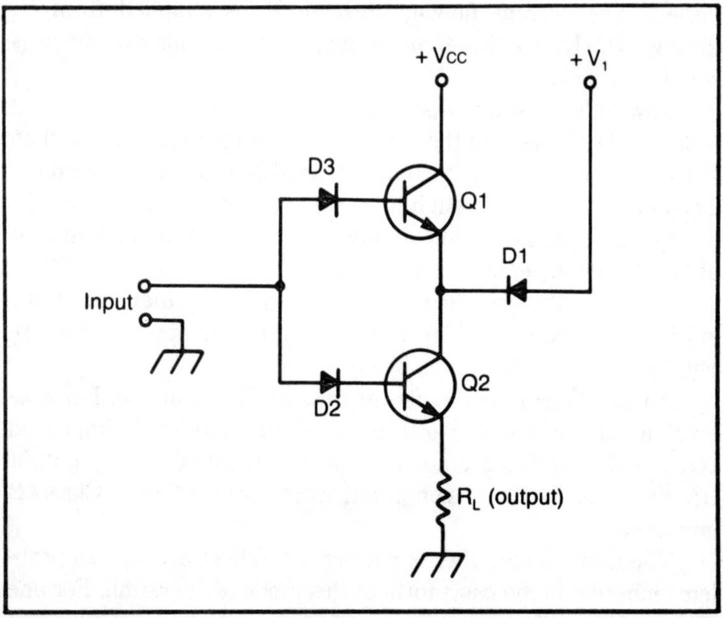

Fig. 2-11. Adding a diode to the basic Class G circuit helps minimize crossover distortion.

A third diode (D_3) is also added here. Since Q_2 is reverse biased when the signal voltage is lower than V_1, the base/emitter junction of Q_2 must be able to stand a reverse voltage higher than V_1. Since the maximum inverse voltage of the base/emitter circuit of most transistors is generally low, D_3 is added to the circuit to prevent the flow of excessive reverse current through the base/emitter junction of Q_2. This protects the transistor against the reverse voltage.

Class H Amplifiers

The Class H amplifier is not dissimilar in concept to the Class G amplifier. This mode of amplifier operation was developed by Soundcraftsman of Santa Ana, CA. The basic Class H circuit is shown in Fig. 2-12.

Essentially, the Class H amplifier appears to work exactly like a conventional Class AB amplifier at low power outputs, but, as the signal level approaches the low voltage supply limit, the circuit starts to function differently.

In a standard amplifier circuit, clipping of the signal sets in as the input level increases. In the Class H amplifier special vari-proportional logic-control circuits anticipate the signal's approach to the supply voltage level and start to increase the B+ and B− voltages to allow for additional head room or power output. This action can be continued as much as is needed, until the vari-proportional system reaches its maximum level (the limit of the high-voltage supply). Any attempt to drive the amplifier past this point causes clipping at the high voltage supply level.

Essentially, the Class H amplifier employs logic circuits in an otherwise standard Class AB push-pull amplifier. Normal or average levels are reproduced as in any Class AB amplifier. Most of the time, the Class H amplifier operates at a lower voltage. This improves efficiency because it reduces output-stage dissipation. Output-stage dissipation is directly proportional to the voltages applied across the output transistors.

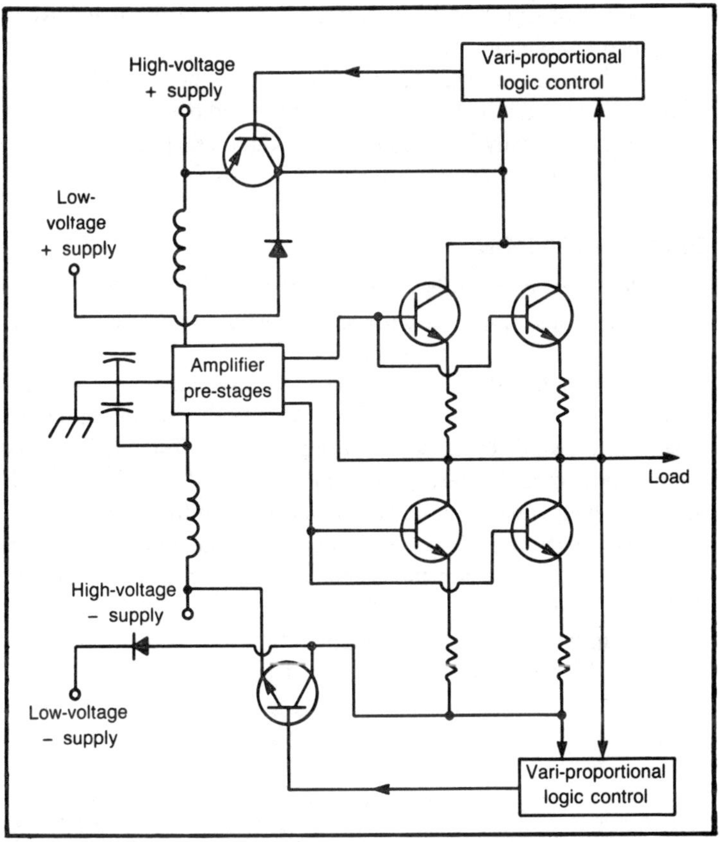

Fig. 2-12. Basic Class H amplifier.

There is even some degree of energy conservation under high powered sine wave conditions too. While the high voltage supply is being turned on to its maximum level, it only reaches that maximum for a very brief period at the peak of the output sine wave. Since the turn on gain of the upper supply voltage is greater than the amplifier gain, the sine wave signal can never catch up with the higher supply (until it reaches final, upper clipping).

According to the manufacturer the inherent slew rate of the higher voltage supply is greater than that of the audio amplifier circuits themselves (approximately 50 volts per μS). The vari-proportional supply logic anticipates the rising waveshape regardless of its shape, and turns the higher supply on with a gain and slew rate exceeding those of the amplifier itself.

The Class H amplifier's chief advantage claimed over other similar efficiency-improving systems is that there is no switching or changing of signal paths within the basic amplifier itself. All of the controls adjusting the power-output capability as needed act only within the power supply circuitry. These controls are completely outside the feedback loop of the audio amplifier circuits, so they theoretically have no effect on the distortion, stability or slew rate of the basic amplifier circuit.

BIAMPLIFICATION

Selecting the amplifier class is not the only approach to coming up with the best trade-off between distortion and efficiency. Another approach is biamplification.

In a biamplification system, the audio signal is fed, as a whole, to one or more filters. At the filter outputs the audio bandwidth is split into two discrete bands, a band of high frequencies and a band of low frequencies. Each frequency band is independently amplified by its own push-pull amplifier. Each amplifier is designed to reproduce its specific band of frequencies most efficiently. The crossover frequency is usually chosen to be somewhere between 400 and 800 Hz.

A similar system is triamplification, in which the audio spectrum is divided into three distinct bands.

Chapter 3

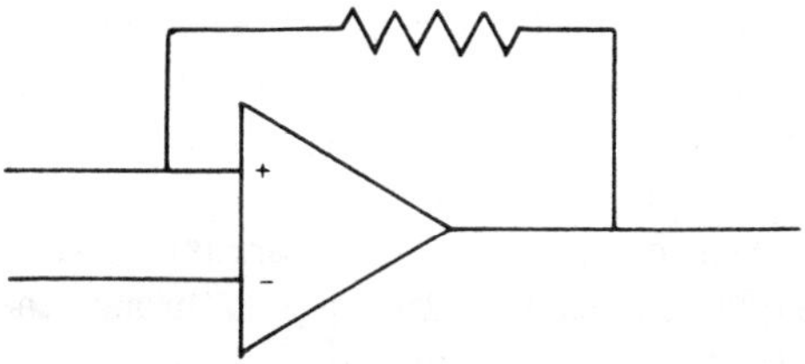

Increasing Gain

THE GAIN OF ANY SINGLE AMPLIFIER STAGE IS LIMITED BY THE beta of the transistor used. While this may be sufficient for some applications, many other applications require more power than a single transistor can put out. This is the case for the majority of audio and rf circuits.

The solution is to cascade multiple amplifier stages. The output from one amplifier stage must be fed to one or more subsequent stages to obtain sufficient gain. When amplifier stages are cascaded in this fashion, the voltage gain of the first stage is multiplied by the voltage gain of the succeeding stages to determine the overall gain of the circuit as a whole.

INTERSTAGE COUPLING

If you are going to construct a multi-stage amplifier, you need some method for interconnecting, or coupling, the various stages. There are several popular approaches.

This discussion concentrates on bipolar transistors. In many applications, FETs might be preferable, because they are not quite as susceptible to some of the problems discussed here. If you can design a workable bipolar transistor circuit, you should have no problems with a comparable FET circuit.

Transformer Coupling

Historically, transformer coupling has been used more

frequently than any other method. Until fairly recently, most audio amplifiers used transformer coupling, particularly in the power output stages. (See the section on output coupling later in this chapter.) Transformer coupling is still used in some audio amplifiers, but it is more common in i-f and rf circuits.

Transformer coupling has enjoyed its popularity because the transformer has several characteristics that make it quite suitable as a coupling device. If you are familiar with basic electronics, you know that the basic transformer is made up of two (or sometimes more) coils of wire wound around a common magnetic-core. The primary coil has N_1 turns of wire, while the secondary is comprised of N_2 turns of wire.

If you place a dc current across the primary, a steady magnetic field is induced in the primary and coupled through the core to the secondary. However, the current in the primary is not induced into the secondary.

On the other hand, when an ac current is fed to the primary, an ac current is induced into the secondary. Assuming an ideal transformer, the induced signal in the secondary coil has the same frequency and waveshape as the input signal across the primary coil.

In an ideal transformer with perfect magnetic coupling between the two coils, the induced current in the secondary (I_2) due to the input current (I_1) in the primary is inversely proportional to the ratio of turns in the coils, or:

$$\frac{I_2}{I_1} = \frac{N_1}{N_2}$$

Similarly, the ratio of the voltages across the two coils is directly proportional to the turns ratio, or:

$$\frac{V_1}{V_2} = \frac{N_1}{N_2}$$

According to Ohm's law, the impedance (Z) is equal to the product of the voltage (V) and the current (I). This allows you to multiply these equations to discover that the impedance ratio is proportional to the turns ratio squared, or:

$$Z = \frac{N_1^2}{N_2^2}$$

In other words, this equation tells us that the impedance of the secondary (Z_2) appears reflected into the primary as an impedance (Z_1) equal to Z_2 multiplied by the square of the turns ratio $((N1/N2)^2)$.

Transformers work well for inter-stage coupling at rf frequencies, but in the audio range, suitable transformers are generally rather expensive and bulky. For this reason, RC coupling is more widely used in audio applications today. Transformer coupling is still widely used in modern rf circuits.

RC Coupling

Multiple amplifier stages can also be coupled through a resistor and a capacitor (an RC network). In this method of coupling, a capacitor is used to couple a signal from its source to the input of an amplifier in order to isolate the source from the bias circuit. If the source and the bias circuit are not isolated in this fashion, the bias on the transistor could be adversely altered by the signal source. That undesirable situation can occur in a two-stage (or more) coupled circuit as well as anywhere else, but in a multi-stage amplifier, there is still another factor to consider. In this type of circuit, the collector voltage of the first transistor can also affect the base current of the second. By the same token, the collector voltage of the first stage can be affected by the current drawn by the base circuit of the second stage.

However, this kind of inter-stage interaction can be accounted for by the circuit designer, so a coupling capacitor between the two stages is not always required. If there is no coupling capacitor, we have what is known as a direct-coupled amplifier.

Direct Coupling

The chief advantage of direct coupling is that it offers improved low frequency response. (Remember that a capacitor acts as a high impedance to low frequencies.)

Another problem inherent in RC coupling is that capacitor-resistor circuits introduce phase shifts. Such circuits tend not to be very linear and so they often add distortion. Because of these problems, direct coupling has been an important goal for designers of high fidelity equipment.

OUTPUT COUPLING

Besides inter-stage coupling, you must also consider how the amplifier's output is coupled to the actual output device (usually a loudspeaker). Since most loudspeakers are low impedance devices (typically 4 to 16 ohms), and most transistor circuits have a much higher output impedance, careful consideration must be given to the coupling method used.

Transformer Output Coupling

Just as transformer coupling was once the norm for inter-stage coupling, many older audio devices also employed impedance-matching transformers to couple the amplifier output to the loudspeaker. Transformer output coupling is still used in much low-end audio equipment today, especially in pocket radios.

In better, high-fidelity equipment, transformers are generally avoided because they can severely limit the fidelity of the signal. Most modern audio amplifiers use one of a variety of types of transformerless circuits to drive the output.

Transformerless Capacitor Output Coupling

In older designs transformerless amplifiers generally employed bipolar power-transistors. Today the trend is increasing use of power VFETs and MOSFETs. These devices are not prone to such common bipolar transistor problems as thermal runaway and breakdown. In addition, FETs are considerably more linear than bipolar transistors, so FET amplifiers tend to offer lower inherent distortion. This means that a FET amplifier requires less feedback. A bipolar transistor amplifier would need more negative feedback to achieve the same final distortion level.

Since FET amplifiers need less feedback, it follows that there are fewer instability problems due to the feedback network.

Finally a FET output stage doesn't require as much impedance matching as a similar bipolar transistor circuit. The output can be fed to the loudspeaker through a simple capacitor. The capacitor's function is to block any dc voltage from the speaker's sensitive voice coil.

Direct Output Coupling

Once again we are faced with the problem of poor low frequency response through the coupling capacitor. Many recent designs use direct output coupling.

There are some potential problems for amplifiers that don't use a capacitor between the output transistors and the loudspeaker. The most significant problem, of course, is that there is no way of keeping dc from flowing through the speaker. A popular solution is for the output devices in the final push-pull stage to be connected to equal positive and negative voltage supplies. The voltage at the junction of the output devices is zero, assuming that equal idling current flows through the two complementary pairs of transistors. This requirement can usually be met with a simple current adjustment.

Unfortunately, this balanced relationship holds at only one temperature. If the temperature rises or falls in the preceding dc coupled stages the idling current at the junction may take on a nonzero value.

To correct this problem, differential amplifiers are often used to drive the output stages. In a differential amplifier, a current change in one of the devices causes an equal and opposite current change in the other device. The overall balance between the devices is maintained.

Chapter 4

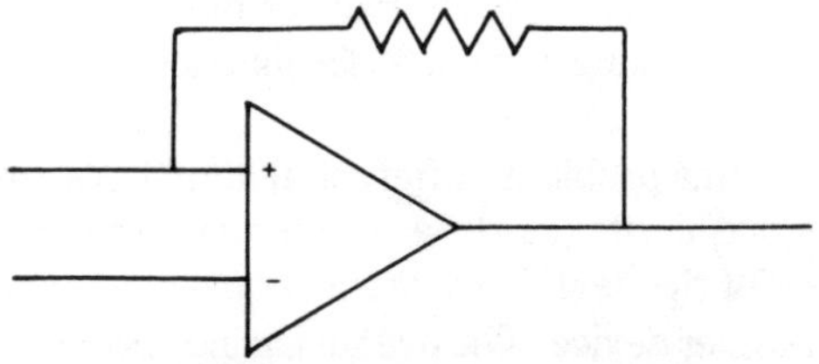

Amplifier Problems

AN IDEAL AMPLIFIER REPRODUCES ANY SIGNAL PASSING through it exactly. The output signal is precisely the same as the input signal, except for the amplitude and possibly the impedance. (In some cases only the impedance is changed—buffer amplifiers do not change the signal amplitude.) Everything applied to the input of the amplifier should appear at the output. Similarly, the output signal should contain no extra components that were not present in the original input signal.

Real amplifiers are susceptible to a number of potential problems. In some cases the effects are negligible in the specific application at hand. In other applications, additional circuitry may need to be added to the system to compensate for certain shortcomings of the amplifier.

DISTORTION

The most obvious type of problem in an amplifier's performance is distortion. In a nutshell, distortion is any effect that alters the waveshape of the signal.

There are several basic types of distortion. The most important types are discussed in the next few pages.

THD

In most amplifiers, the most significant type of distortion is THD, or total harmonic distortion.

Most ac signals (and sounds) are made up of multiple frequency components. Only the sine wave has just a single frequency component. The basic repetition rate of the waveshape is called the fundamental. Additional frequency components that are higher than the fundamental are called overtones. (Occasionally there is an additional frequency component that is lower than the fundamental. This is called an undertone, or subtone. Such frequency components are relatively rare.

If the overtone is an exact integer multiple (2X, 3X, 4X, etc.) of the fundamental frequency, it is called a harmonic. Most periodic waveforms are made up primarily of a fundamental and one or more of its harmonics. Not all waveshapes contain all of the possible harmonics. For example, any symmetrical waveshape (such as a triangle wave, or a square wave) contains no even harmonics (2X, 4X, 6X, etc.).

Non-linearities in an amplifier circuit can cause it to create false harmonics of any signal passing through it. The output signal may contain frequency components that were not present in the input signal. This is called harmonic distortion. THD is a measurement of all such false harmonics generated by the amplifier. (It seems that false low-order harmonics tend to be somewhat less objectionable than false high-order harmonics of an equal amplitude.)

The carrier storage effect of bipolar transistors is one source of non-linearities that can lead to serious harmonic distortion. One solution to this problem is to use FETs in place of bipolar transistors. As mentioned back in Chapter 1, FETs are not prone to the carrier storage effect. This is particularly true of the Vertical-FET, or V-FET, which is highly linear. By using a V-FET, false high-order harmonics are virtually eliminated, or at least minimized. Since the V-FET is so linear, only relatively small amounts of negative feedback are needed for a wideband frequency response and a high degree of stability.

Intermodulation Distortion

Another common type of distortion is known as intermodulation (IM) distortion. In most practical applications, more than a single signal passes through the amplifier at the same time. For example, in music, two or more instruments are playing simultaneously. If these multiple signals interact, the result is intermodulation distortion.

To give you a better idea of what happens in intermodulation distortion, let's consider what happens when an amplifier simultaneously passes two simple sine wave signals. Signal A is a 800 Hz sine wave, and signal B is a 1200 Hz sine wave.

Ignoring the effects of harmonic distortion, the output signal should be made up of just the two original frequency components:

800 Hz
1200 Hz

But with intermodulation distortion, these frequency components can interact. The output includes phantom frequency components at the sum and difference of the original signal frequencies:

- ☐ 400 Hz—DIFFERENCE (1200 - 800)
- ☐ 800 Hz—ORIGINAL SIGNAL
- ☐ 1200 Hz—ORIGINAL SIGNAL
- ☐ 2000 Hz—SUM (800 + 1200)

If more complex signals are included in the input, each of the frequency components behaves as an individual sine wave, generating sum and difference frequencies with each of the other frequency components.

Obviously, if intermodulation distortion is severe, the output could end up a real mess for even a moderately complex musical passage. Fortunately, this type of distortion is generally not terribly severe. The phantom (sum and difference) frequency components are relatively weak in amplitude. Even so, great care must be taken in designing a high fidelity audio amplifier to prevent intermodulation distortion from reaching an annoyance level that would be unacceptable in audiophile applications.

Crossover Distortion

In earlier chapters we mentioned that push-pull amplifier circuits are susceptible to crossover distortion, which is a break in the waveshape at the point when one transistor switches off and the other switches on. This is also sometimes called notch distortion because of the appearance of the distorted waveform. Notch distortion is a problem especially with bipolar transistors.

A common cause for this type of distortion is improper biasing of a Class B transistor circuit. Even with properly biased output stages there can be a noticeable amount of notch distortion, primarily because of the carrier-storage effect which was discussed earlier.

While most other forms of distortion vary with the power output level, crossover distortion is present at all power output levels. As a result, this type of distortion tends to be more audibly objectionable at low listening levels. When the desired signal has a low amplitude, the distortion obviously constitutes a greater percentage of the total reproduced signal. At higher listening levels the same level of distortion is more completely masked by the desired signal. Since a V-FET is devoid of any carrier-storage effect, it is not subject to notch distortion.

The advantages of V-FETs have been mentioned several times already. You may be wondering why bipolar transistors are ever used in amplifier circuits at all. One reason is that in older designs, V-FETs were not a viable choice. They still tend to be considerably more expensive than comparable bipolar transistors, especially when high power levels are required. High power FETs are still a fairly new development. Undoubtedly V-FETs will become more and more common in new high fidelity equipment.

TIM Distortion

For some years electronics engineers were faced with a rather confusing problem. Two amplifiers with identical specs and the same distortion levels might sound very different from each other. The only reasonable answer was that there must be some other form of distortion that was not yet accounted for.

In the mid to late 1970s, the audio industry started to become aware of transient intermodulation (TIM) distortion. TIM distortion is found in an amplifier that has a large amount of negative feedback in its main feedback loop and a certain amount of time or phase delay between the input and output signals. Essentially, when a very fast musical transient or pulse is fed to such an amplifier, the feedback needed to reduce the amplitude of that transient signal at the input arrives too late. The amplifier is overloaded and/or momentary clipping occurs. Other program signal elements are also distorted or even obliterated by this clipping.

TIM distortion is trickier to measure than other types of distortion like THD and IM. This is because we can't measure TIM distortion with the standard steady-state sine wave input signals normally used for distortion measurements. Even with an amplifier with severe TIM distortion, such a steady-state signal results in a composite output signal that merely has some moderate phase shift rather than a momentary overload or overshoot.

Many experts believe TIM distortion is the main reason for the perceived differences in sound between similarly rated tube and transistor amplifiers. Even the very best of the old tube amplifiers never had the low THD and IM ratings claimed by newer solid-state equipment, but many people feel the tube amplifiers sounded better or "warmer."

TIM distortion may be more significant than the difference in transfer characteristics between tubes and linear solid-state devices. Today's circuit designers can make transistor amplifiers, which, when tested with steady state signals at least, display exactly the same transfer characteristics and overload waveforms as did the earlier tube amplifiers. But the solid-state versions still sound different.

Recent research indicates that there does seem to be a strong correlation between a high level of TIM distortion and the vaguely harsh sound quality that has been attributing to certain amplifier designs for many years. Transistorized amplifiers tend to be more prone to TIM distortion because until recently they were generally designed with more feedback than most earlier tube designs.

Reactive components (such as capacitors and inductors) cause the feedback signal to be subjected to a finite time delay. Further delay results from the transit time of the amplifying devices themselves. In effect, the feedback signal arrives back at the input somewhat delayed in time. If a pure sine wave is fed to an amplifier and the delay amounts to as much as 45 degrees of lag between the input signal and feedback signal, the net signal is still perfectly sinusoidal in shape.

Negative feedback is shifted by 180 degrees. If you add 45 degrees of delay, you find that the feedback signal is phase shifted 225 degrees. The net input signal is now reduced by a little less than 6 dB, but it is still exactly sinusoidal in shape. The output signal is somewhat displaced in time from the original input, but this is rarely significant.

What if the input signal is not a sine wave, but a step-function such as a squarewave? As a rule, musical signals are closer to such steep rising functions than to sine waves, especially when the music contains a lot of transient information or fast instrument attack times.

Figure 4-1 illustrates the effects of TIM distortion on a square wave, assuming that the circuit and feedback delay are the same

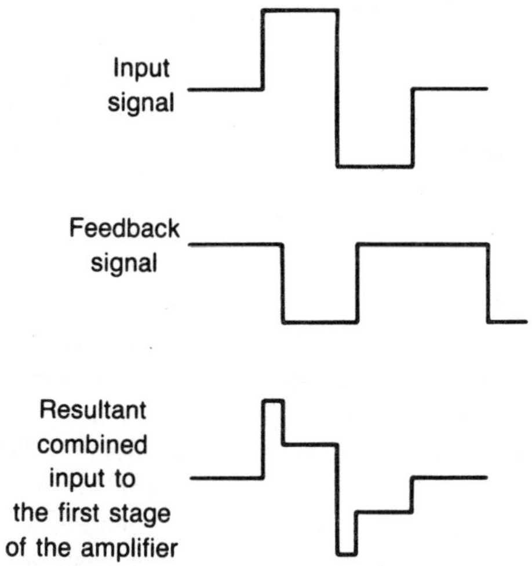

Fig. 4-1. The effects of TIM distortion on a square wave.

as in the previous example. Here we see that the net input amplitude actually increases in the positive going direction by 6 dB for the first eighth of the cycle, thanks to the step function nature of the original waveform. The instantaneous amplitude of the feedback signal is in phase with the input signal and adds to it rather than subtracting from it.

Even if the step function at the input is a brief, one-time event (common in music) the desired input signal amplitude reduction that the negative feedback should have caused won't take place, even though the net input amplitude might not have increased initially. This is due to the time-delayed feedback.

We specified 6 dB of feedback in the examples. This is actually a rather small amount. In most practical amplifiers, the loop feedback is 40 dB or even more. For a steady state signal that drives an amplifier to, say, its rated output of 20 watts, removing the required 40 dB of feedback for however short a time period theoretically requires that the same amplifier produce an instantaneous peak power output of 200,000 watts, obviously beyond the circuit's actual capability. The result is severe clipping and distortion.

Matters get even more complex when listening to real music, rather than nice, neat test tones. In the real world we aren't listening to clean, simple step-function signals, but with complex signals in which multiple step-functions may be mixed with other sinusoidal signals or step-functions at other frequencies.

This all boils down to the unfortunate fact that getting a true measurement of an amplifier's TIM distortion level is no easy task. However, there are some available techniques for getting at least an approximate reading.

One method of measuring TIM effects is to use a compound input signal consisting of a 500 Hz squarewave mixed with a 6000 Hz tone whose amplitude is one fourth or one fifth that of the lower frequency squarewave signal. This TIM test signal is applied to an amplifier and the levels are adjusted so that the peak power output is slightly lower than the amplifier's maximum continuous sine wave power output level. If TIM distortion is present, the first cycle of the superimposed high frequency signal, is blurred by the momentary absence of the properly out-of-phase feedback during the fast risetime and falltime of the low frequency squarewave. This blurring occurs because the amplifier has been driven into clipping.

The effects of TIM distortion can be viewed by another method, using a spectrum analyzer. The same basic squarewave/ sine wave composite signal is used as a test signal fed to the amplifier. Any extraneous frequency components in the output indicate TIM distortion. Unfortunately, this type of measurement does not easily lend itself to numerical interpretation.

CROSSTALK

Additional problems show up in a stereo amplifier. Essentially, a stereo amplifier is nothing more than two separate amplifiers used in unison. One amplifier handles the sound for the right speaker, while the other takes care of the left one. In theory, the amplification of the two channels should be completely independent. What occurs in one channel should have absolutely no bearing on the other channel.

If the right and left amplifiers share any common circuitry at all, this ideal is not achieved. Some of the right channel signal leaks over into the left channel amplifier, and vice versa. This inter-channel bleedthrough is called crosstalk.

Fortunately it is not really necessary for an audio amplifier to be completely devoid of crosstalk, since all stereo recording media have some degree of crosstalk anyway. For an absolute minimum of crosstalk, the amplifiers for each channel should be 100 percent independent. Many high-end stereo systems incorporate separate power supplies for each channel. Signals can leak from one channel to another through power supply connections, although in many cases the importance of this is questionable, since the recorded media being played through the amplifier may have a considerably higher amount of crosstalk itself.

A related problem is dynamic crosstalk. Here, the low frequency content of one channel modulates the signal in the other channel. While dynamic crosstalk is measurable, the audibility of such effects is questionable. If the effect cannot be heard, its importance is clearly negligible in an audio amplifier.

THERMAL PROBLEMS

All electronic components are heat-sensitive to some degree. This means that their values and/or certain operating parameters may change with fluctuations in temperature. In addition, some components, especially solid-state devices, can be damaged or destroyed by excessive heat.

At the same time, all electronic circuits generate heat. Of course, transistors don't generate as much heat as tubes, but they can still run quite hot. In many high power circuits, some sort of heat sinking, or even a cooling fan proves necessary.

As a rule, changes in the ambient (room) temperature generally don't cause any noticeable changes in the operation of an amplifier. However, temperature-related problems can show up due to heat generated by the circuitry itself.

If insufficient heat sinking or ventilation is provided, semiconductor components can self-destruct from their own operating heat. Commercial electronic equipment always has ventilation slots in the cabinet. These openings must never be blocked off while the unit is being operated. Heat within the cabinet can quickly build up to unacceptable and damaging levels. Any mounting arrangement must permit adequate ventilation. Some stereo system display cabinets do not take this into account, and such models should be avoided. Care must be exercised when stacking pieces of equipment. Many devices have their ventilation slots on the top or the bottom of the cabinet. These units should not be stacked so that any of the ventilation slots are blocked by another piece of equipment. Also watch out for arrangements in which one piece of equipment could exchange its internal heat with a second.

Even with adequate ventilation and heat sinking, bipolar transistors can run into serious heat problems. As a bipolar transistor conducts more current, it gets hotter. At the same time, the hotter a bipolar transistor gets, the more current it conducts. This can lead to a dangerous loop condition known as thermal runaway. If the transistor starts to conduct too much current (perhaps due to a slight short), its temperature will rise, causing it to conduct still more current, causing it to get even hotter. This cycle continues until the semiconductor crystal self-destructs from its own generated heat. In many better circuits using bipolar transistors, special protection circuitry is included to avoid thermal runaway.

FETs, on the other hand are not prone to thermal runaway, because they exhibit an inverse relationship between temperature and current flow. As the temperature increases, the current flow decreases. In a sense, FETs in a power amplifier can be considered self-protecting.

POSITIVE FEEDBACK

Earlier we saw how negative feedback, in which the output is 180 degrees out of phase with the input, can improve an amplifier's linearity. Positive feedback, however, is a whole different story. In positive feedback the output and the input are in phase with each other so they add. The signal is amplified, fed back to the input, then amplified again. This continues in an infinite loop until the amplifier breaks into oscillation. The amplifier becomes an oscillator. (A basic oscillator circuit is nothing more than an amplifier stage with frequency specific positive feedback.)

Positive feedback is what leads to that ear-piercing squeal in a PA system when the microphone is too close to the loudspeaker.

Chapter 5

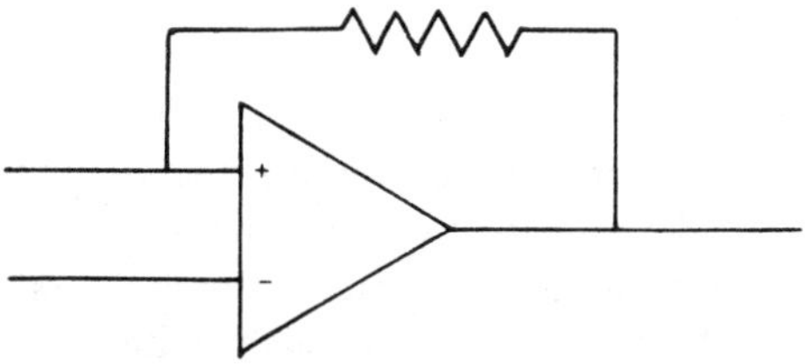

Audio Amplifiers

AUDIO AMPLIFIERS HAVE BEEN MENTIONED SEVERAL TIMES in the earlier chapters. This chapter examines the specifics of amplifying audio signals, primarily in a high fidelity (stereo) system.

THE BASIC AUDIO AMPLIFIER

While there are often major differences between audio amplifiers from different manufacturers (or sometimes even between different models from a single manufacturer), there are also important similarities.

Figure 5-1 shows a block diagram for a typical stereo amplifier. This basic three stage audio amplifier chain has been widely used in high-fidelity equipment over the years. The three primary stages are:

- ☐ Preamplifier
- ☐ Driver
- ☐ Power amplifier

Most audio amplifiers follow this basic pattern, although there are some occasional exceptions.

Each stage is designed to serve a particular purpose. The preamplifier boosts the voltage level of the input signal to a level usable by the driver. The driver then increases the power level of

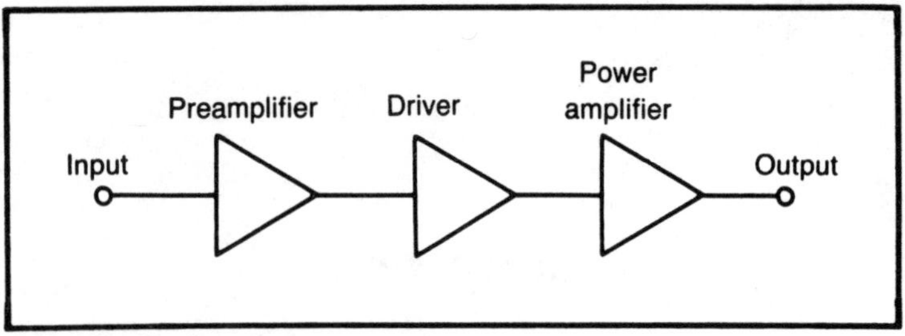

Fig. 5-1. Basic audio amplifier chain.

the amplified signal enough to drive the output power amplifier stage. Finally, the output power amplifier develops the power to drive the output device, usually a loudspeaker.

The simple circuit shown in Fig. 5-2 is found in a number of automobile and home radios, but it isn't quite suitable for high fidelity applications. This circuit is a single-ended Class A amplifier. Output impedance matching is accomplished via a choke or autotransformer.

While very simple, this circuit has some significant disadvantages. Since this is a Class A amplifier, the output collector current flows all of the time, even when there is no input signal. This means the circuit has low efficiency and a lot of heat is generated. To protect against excessive buildup, a 3 to 5 watt fused resistor (fusistor) is often placed in series with the transistor.

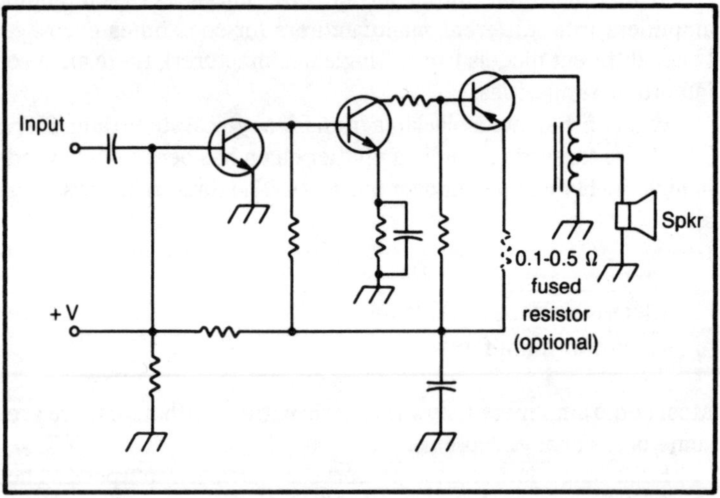

Fig. 5-2. A simple amplifier circuit found in a number of automobile and home radios.

Even though Class A circuits are linear, the actual fidelity of this circuit is not particularly good unless sufficient negative feedback is provided. A small amount of negative feedback results from the presence of the protective fusistor, but this is rarely sufficient and an additional negative feedback network must be provided in most cases.

Most audio circuits use one of two basic types of feedback. The first is known as the "second collector to first emitter" system. As Fig. 5-3 shows, the name is descriptive. By selecting the proper component values, this feedback circuit can convert a relatively mediocre amplifier into a reasonably high fidelity unit.

The second common feedback circuit (shown in Fig. 5-4) is known as the "second emitter to first base" circuit. Often, the feedback path consists of just a single resistor.

Most practical audio amplifiers use some form of the push-pull circuit. The basic transistor push-pull circuit is illustrated in Fig. 5-5. The push-pull circuit is generally preferred over other types because of its power handling capability and overall fidelity. This type of circuitry is found in virtually every audio application, from $5 pocket radios to relatively high-priced, medium grade console radios and stereos.

In recent years, the basic push-pull circuit has lost some ground.

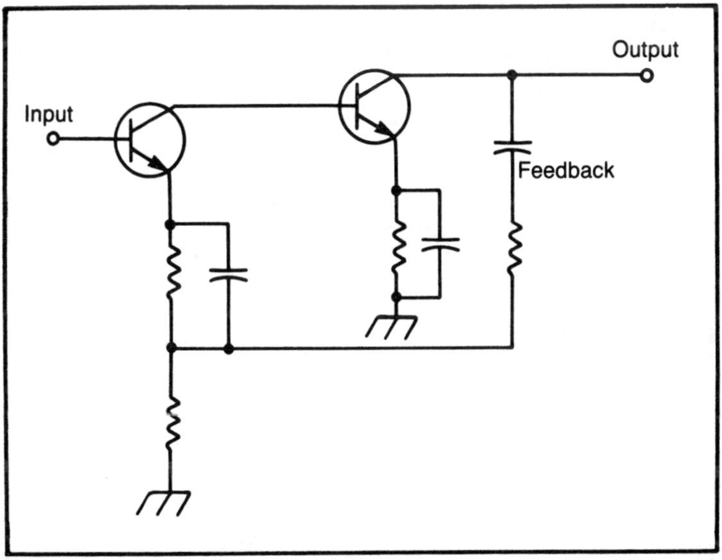

Fig. 5-3. One popular form of feedback is second collector to first emitter feedback.

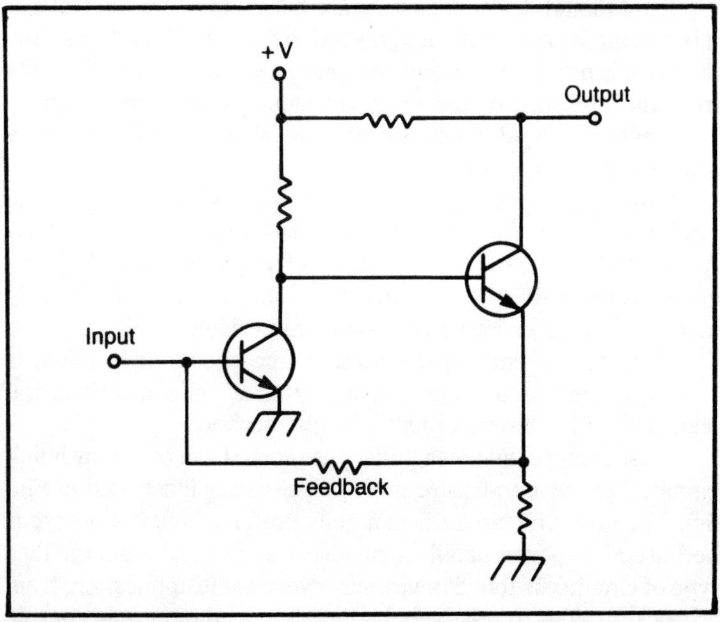

Fig. 5-4. Another common form of feedback is second collector to first base feedback.

It is less cost-efficient than certain other circuits of more recent design.

A variant on the basic push-pull amplifier is the split-secondary, totem-pole circuit, shown in Fig. 5-6. This type of circuit is found mostly in imported radios, although it does show up in some domestic models. This circuit can be quickly recognized by the se-

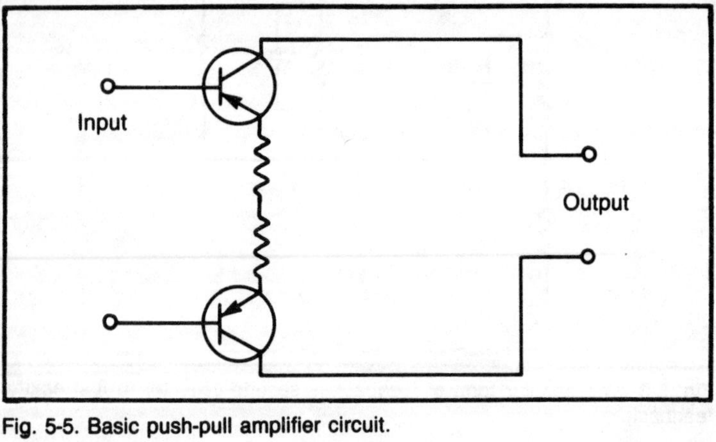

Fig. 5-5. Basic push-pull amplifier circuit.

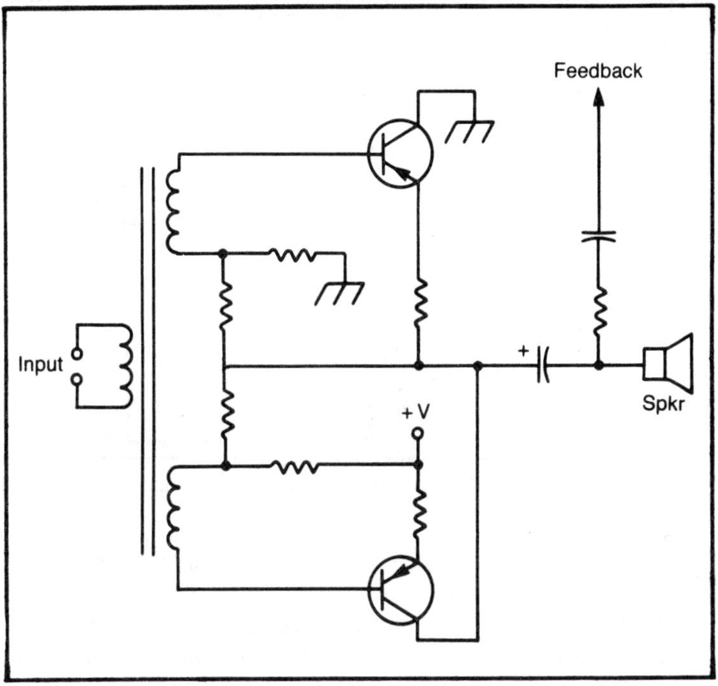

Fig. 5-6. A variation on the basic push-pull amplifier: the split-secondary, totem-pole circuit.

ries connection of the output transistors and the split-secondary interstage transformer (T1).

All push-pull amplifier circuits share one important factor in common: the need for phase-splitting the input signal to create two signals that are 180 degrees out-of-phase to drive the two halves of the push-pull circuit. A center-tapped transformer was often used for this purpose in many older designs. This technique is illustrated in Fig. 5-7. Sometimes a split-secondary interstage transformer was used, as illustrated in Fig. 5-8.

Many modern circuits take a different approach to splitting the input signal. A transistor phase inverter (Fig. 5-9) is often used in place of the interstage transformer. One driving signal is taken from the transistor's collector, while the other driving signal is taken from the emitter.

A special IC preamplifier that has both inverted and noninverted output terminals can also be used to provide drive signals of opposite polarity. These ICs put out wideband push-pull outputs from a common input signal.

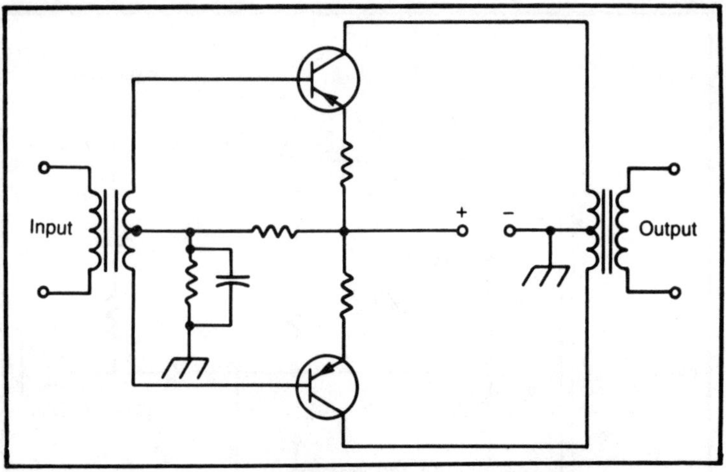

Fig. 5-7. A center-tapped transformer, used for signal splitting in push-pull circuits.

Some push-pull circuits incorporate complementary-symmetry, as illustrated in simplified form in Fig. 5-10. Since pnp and npn bipolar transistors require signals of opposite polarity to perform the same basic functions a single input can be used without a signal splitter. Notice that the speaker is connected to the midpoint of the two series-connected power transistors. An output transformer is not required.

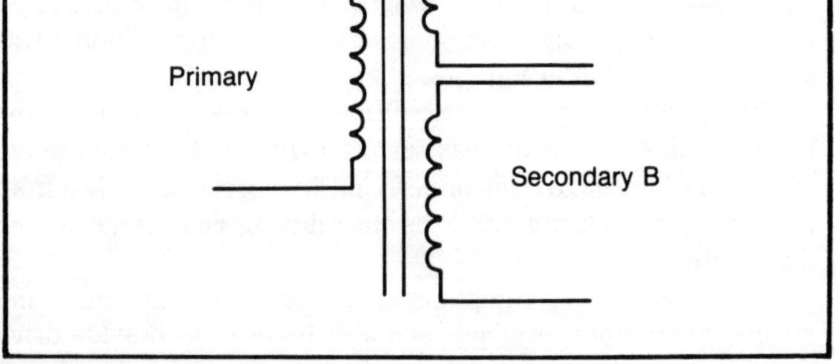

Fig. 5-8. A split-secondary interstage transformer, also used for signal splitting in push-pull circuits.

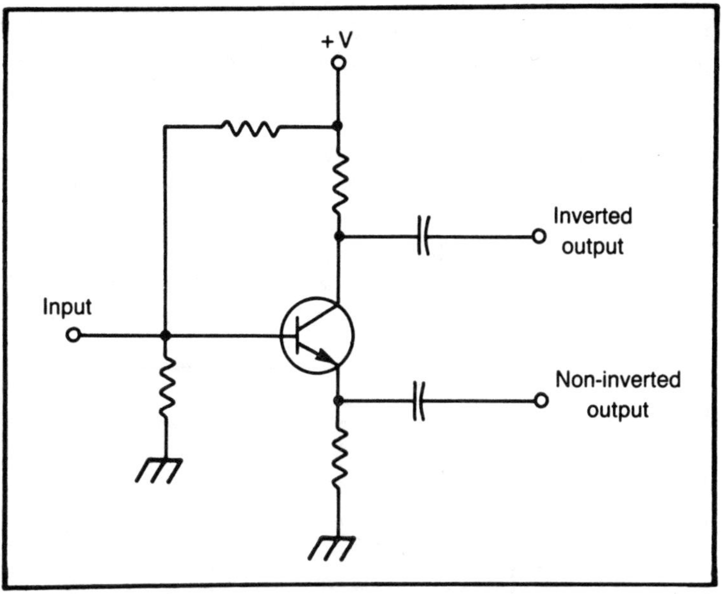

Fig. 5-9. A push-pull circuit using a transistor phase inverter for signal splitting.

The chief reason complementary-symmetry amplifier circuits aren't used more often is that it is difficult to locate matching pnp and npn transistors. If we examine the manufacturers' spec sheets we find that there are actually just a few available choices for

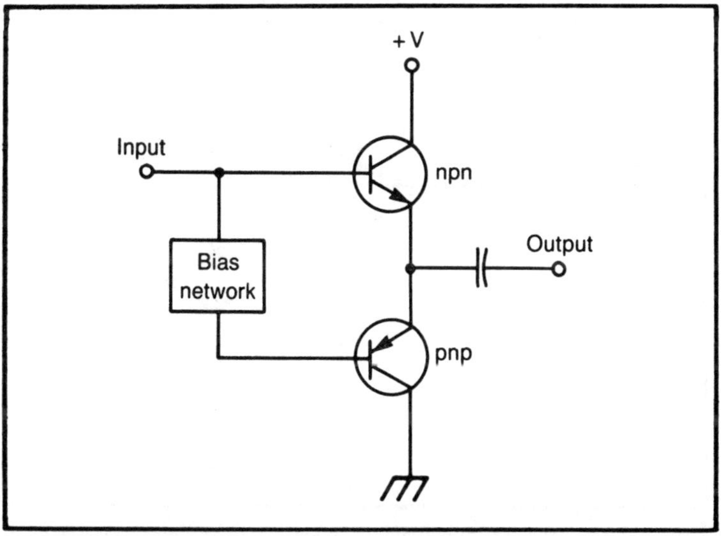

Fig. 5-10. A push-pull circuit based on complementary-symmetry.

matched complementary pairs at any given output power level. This problem gets more severe as the amplifier's power level increases. It generally isn't too difficult to locate suitable matched pairs of transistors for low and medium power complementary circuits. Even for outputs of a few watts, several choices are available. But if we need higher power we may have considerable difficulty finding adequate complementary transistors.

To get around such problems, the basic complementary-symmetry circuit has been modified to create the quasi-complementary circuit. As shown in Fig. 5-11, this circuit uses a

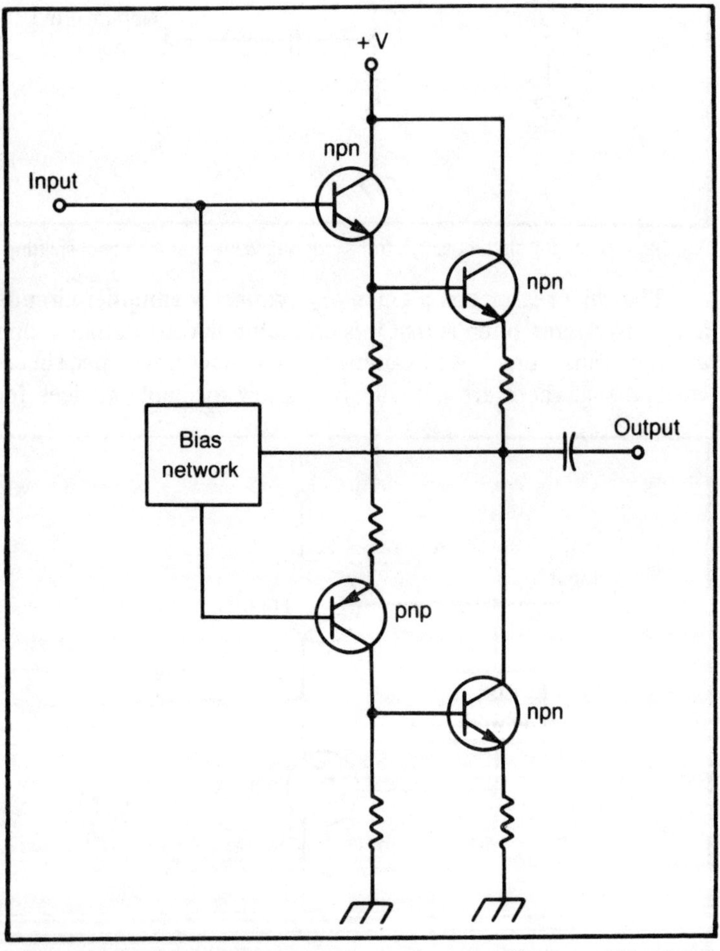

Fig. 5-11. The quasi-complementary circuit, a modification of the basic complementary-symmetry principle.

totem-pole output in which the same type of npn or pnp transistors are in series with each other and complementary driver stage Q1/Q2. There is usually no problem in locating the medium power complementary drivers and matched (identical) output transistors required for this type of circuit.

DEFINING AUDIO AMPLIFIER CLASSES

When dealing with simple, single-stage circuits, there is no problem in defining the amplifier class. But in sophisticated multistage audio amplifiers, various stages may be designed in different class modes. What is the class of the amplifier as a whole?

Traditionally, the industry has resolved this problem by emphasizing the class of the output stage. This is reasonable, since the final stage is the one that actually delivers power to the speakers.

That is simple enough, but several important amplifier characteristics (including frequency response, gain, thermal drift, and slew-rate, among others) are determined by earlier stages, such as the driver or voltage-amplifier stages. The driver stage may also have a significant effect on the distortion characteristics of the amplifier as a whole.

SPECIFICATIONS

Many different characteristics define how an audio amplifier sounds. To allow objective comparisons between various audio amplifiers and related equipment, a number of standardized measurements have been devised. The results of these measurements for a given amplifier are called its specifications, or specs.

The measurement tests are standardized by the Institute of High Fidelity (IHF), and, in some cases the Federal Trade Commission (FTC). Compliance with IHF regulations is purely voluntary, while FTC requirements have the force of law. Compliance with IHF standards is almost universal within the industry. Reputable manufacturers willingly adhere to these standards. Without such standards, utter chaos would reign throughout the industry. There would be no consistent way to compare one model with another.

The first specification most people think of is power. How many watts can the amplifier put out? Actually the wattage race has been greatly exaggerated. Beyond a certain point, the law of diminishing returns sets in.

The human ear does not hear differences in volume in a linear manner. A doubling of perceived volume requires a tenfold increase in power. To get twice the sound of a 10 watt amplifier, you need a 100 watt amplifier. A 35 watt amplifier sounds barely louder than a 10 watt unit, but there might be a considerable increase in the price.

On the other hand, there is certainly some merit in the idea that you should get as much power as you can afford. Most home listening is at a power level of no more than a few watts. Even if you like floor-rattling sound, it doesn't take all that much power to accomplish this in a room of the dimensions found in a typical home. But music contains considerable dynamic variations in amplitude. While the average power requirement may be just a few watts, brief peaks require well over 100 watts to reproduce without clipping. In other words, the amplifier must be able to reproduce sounds cleanly at powers well over the comfortable listening level.

Until fairly recently, it was very difficult to compare the actual power capabilities of various amplifiers. There are many different ways to measure power. Most are equally valid, but incompatible. If amplifier A is rated for 10 watts average, amplifier B is rated for 40 watts rms, and amplifier C is rated for 75 watts peak power, how can you make a reasonable comparison of their relative power levels?

Another confusing factor is that an amplifier's maximum power level is not a clear-cut point. An amplifier can put out more than its rated power, but with increased distortion. The maximum acceptable level of distortion obviously influences the amplifier's maximum usable power level. If amplifier A is rated at 25 watts at 1 percent distortion, and amplifier B is rated at 45 watts at 3 percent distortion, you could be misled if you looked only at the wattage figure. Actually, in this example, amplifier A is probably the more powerful unit.

Many manufacturers in the past didn't bother mentioning which type of wattage they were measuring, or the distortion level. Such specifications were utterly useless for comparing various amplifiers.

Eventually, the FIC stepped in and introduced a set of stringent rules standardizing hi-fi amplifier power ratings. It is now the law that audio amplifiers must be rated on the basis of their ability to deliver continuous sine wave power into a specified resistive load. The amplifier does this over a specified band of audio frequencies at some specified value of THD.

While these regulations have been a boon for the consumer for the most part, they have not been a completely unmixed blessing. To some extent they have actually complicated the problem of interfacing amplifiers with loudspeakers. Generally, you do not listen to continuous sine waves. In actual operation, an amplifier might handle the more complex waveforms found in actual program sources quite differently than the continuous sine wave signals employed in the standardized tests. In some cases, the results of the tests may not correspond very well to actual listening conditions.

Fortunately, the test errors tend to be fairly consistent and can be more or less ignored in most comparisons. The continuous sine wave power reading tends to be on the low side. Most amplifiers can put out more short-term power when fed with typically transient musical signals than it can when amplifying continuous sine wave signals.

Additional problems crop up when you try to match up amplifier and speaker power ratings. A 50 watt per channel stereo amplifier, connected to a speaker system that is nominally rated to handle 50 watts, may well be driven to beyond its power-handling capability by transient signal peaks. These transient peaks, though instantaneously well in excess of 50 watts may not cause amplifier clipping because they last such a short time. But when such excessive power levels are fed to the speaker, even intermittently, the speaker can be overloaded and damaged. At the very least, the speaker cone will probably be driven into its non-linear operating region, resulting in distortion (from the speakers, rather than from the amplifier).

In addition, the nominal power handling capability rating of a speaker should be taken with a hefty grain of salt. This rating might be based upon the capabilities of the heavier-duty woofer, while the power handling capacity of the mid-range or high-frequency drivers in the system are likely to be considerably less. If very strong high frequency signals are fed to such speaker systems (perhaps due to super-audible amplifier oscillation, amplifier-generated harmonics due to inadvertent amplifier clipping, or non-typical frequency distribution of musical energy commonly encountered with some forms of electronic music), the tweeter or mid-range speaker can be blown, even though the overall power handling capability of the system is not exceeded. The woofer is left undamaged under these circumstances.

The IHF standards set down in 1960 were rendered obsolete by the FTC's ruling in 1974. The IHF responded with a new, more extensive set of audio amplifier specifications and testing procedures ("Methods of Measurement for Audio Amplifiers," IHF-A-202, 1978).

The new IHF specifications include a measurement of dynamic headroom, to help consumers distinguish between two amplifiers with similar continuous power-output ratings, but noticeably different loudness levels when fed with the same short-term dynamic music signals.

A leading factor in determining an audio amplifier's dynamic headroom is the design of the system's power supply. Under actual music listening conditions, an amplifier with a very "stiff" (tightly-regulated) power supply may not be able to produce much more output power than its rated continuous power level. But a similar amplifier with a "softer" (less tightly-regulated) supply may be able to significantly exceed rated continuous power levels for brief periods. The difference between these two amplifiers is summarized by their dynamic headroom ratings, which are given in decibels (dB).

The dynamic headroom rating can vary from 0 dB (for the stiffly-regulated power-supply amplifier) to over +3.0 dB. This new specification represents an attempt to more accurately indicate audible performance under actual listening conditions.

Another type of specification that was often terribly confusing in the past is the S/N (signal to noise) ratio. This characteristic was discussed in Chapter 1. Much of the confusion here has been due to a lack of uniformity in the reference levels used to make the sensitivity and signal to noise ratio measurements.

In the past, an amplifier's input sensitivity was defined as the input signal amplitude required for the amplifier to deliver its rated output (with the volume control turned all the way up to maximum). As an example of the kind of confusion this approach could cause, let's consider two typical audio amplifiers. Amplifier A is rated for 10 watts, while amplifier B has a power rating of 100 watts. Assume that each of these hypothetical amplifiers requires an input signal of 1.0 volt to deliver its rated output. This suggests that the amplifiers have "equal sensitivity," but this is clearly ridiculous. If 1 volt is fed into amplifier B, it sounds twice as loud than if the same input signal is fed to amplifier A. Obviously, the gain of these two amplifiers is not identical, amplifier B is more sensitive than amplifier A, even though the old measurement technique gives no indication of this.

Under the new IHF standards, amplifier sensitivity is still

measured with the volume control turned up full, but the definition has been changed slightly. Now amplifier sensitivity is defined as the voltage required to produce 1.0 watt at the speaker output terminals (or 0.5 volts, in the case of a separate preamplifier), regardless of the full power (or voltage output) rating of the amplifier in question.

Returning to the example using this new standard shows that the input sensitivity of amplifier A (10 watts) is 0.316 volt, while for amplifier B (100 watts) the input sensitivity is 0.1 volt.

Older methods of measuring the S/N ratio were even more confusing. Many of the problems stemmed from the misleading sensitivity ratings.

To understand this problem, let's consider an example. Imagine an integrated amplifier (A) that has a phono input sensitivity rating of 2.0 mV. When the signal is removed and the input jacks are shorted, the signal to noise measurement for amplifier A is 68 dB.

Now, let's take a look at a second integrated amplifier (B), from a different manufacturer. Amplifier B's phono preamplifier S/N ratio is actually the same as that of amplifier A. But this second manufacturer feels that a 68 dB S/N ratio is not a sufficiently impressive spec. If this manufacturer changes the input reference level to 10 mV, the amplifier would overload with the volume control turned up full, as in the sensitivity measurement. (Amplifier B would produce its full rated output from a 2.0 mV input signal with the volume control turned up fully.) To prevent clipping due to overload, the manufacturer reduces the setting of the volume control so that the amplifier's rated power level appears at the output again. In this example, the volume would have to be reduced by approximately 14 dB!

By changing the measurement procedure in this fashion, amplifier B ends up with a signal to noise ratio of 82 dB, but B's phono preamplifier doesn't really offer any improvement over amplifier A, with its S/N rating of 68 dB.

The new IHF standards address this problem and set up a procedure for uniform signal to noise readings. The official S/N testing procedure now requires that the input-signal level be fixed at 5.0 mV for a magnetic phono section and at 0.5 V for a high-level input, such as the tuner, auxiliary, or tape input on a preamplifier or integrated amplifier. At the same time, the output reference must be adjusted (as in the case of sensitivity adjustments) to a 1.0 watt level (for a power amplifier) or 0.5 V (for a preamplifier). This adjustment is made by using the master volume control. A big advantage of this particular requirement is that it calls for a volume control setting that is close to what a consumer might normally use for listening to music.

The IHF has also addressed the choice of the weighting network question that lead to confusing S/N measurements in the past. (See Chapter 1.) Previously, some manufacturers used a weighting network in these measurements, while others did not. In addition, there were several semi-standard weightings to choose from. Tests made with different weighting networks obviously don't give the same results. The most popular weighting networks were discussed in Chapter 1.

The new IHF standard calls for the use of A weighting. The weighting network is placed between the amplifier's output and the level meter used to read the results of the test.

To measure the S/N ratio of a phono preamplifier, most manufacturers in the past used a shorting plug in the phono input jacks to create a no-signal test condition. In many practical phono-preamplifier circuits, this technique does not reflect what actually happens when a magnetic cartridge is connected to those same phono inputs. A magnetic cartridge has a certain amount of dc resistance and a finite amount of inductance. The new IHF standard requires the use of a network that approximates the complex impedance seen by the phono inputs when a cartridge is actually connected in place of the simpler shorting plug. This network is shown in Fig. 5-12.

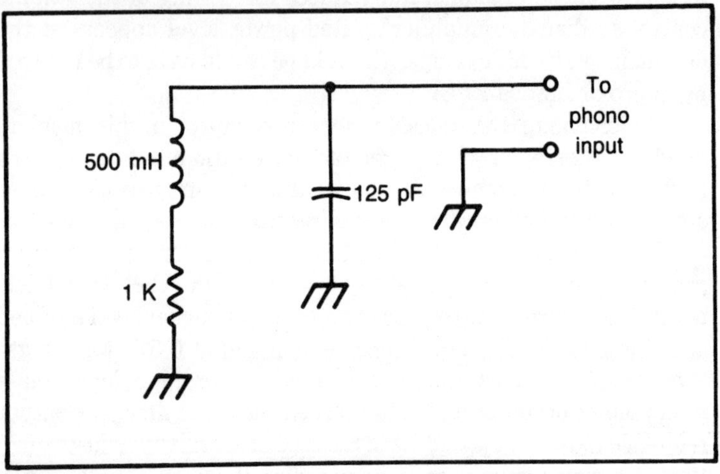

Fig. 5-12. True no-signal tests are best made with a dummy network simulating the characteristics of a magnetic cartridge.

A 1 K resistor is used to terminate the input jacks when S/N measurements of the high level inputs on an amplifier or preamplifier are made.

Some of the new IHF specifications are mandatory: IHF members must make these specs available to consumers. For power amplifiers, these specifications are:

- ☐ Continuous average power output
- ☐ Dynamic headroom
- ☐ Frequency response
- ☐ THD
- ☐ Sensitivity
- ☐ A-weighted signal to noise ratio

For preamplifiers, the specifications are:

- ☐ Frequency response
- ☐ Maximum voltage output
- ☐ THD
- ☐ Sensitivity
- ☐ A-weighted signal to noise ratio
- ☐ Maximum input signal
- ☐ Input impedance

For an integrated amplifier (containing both a preamplifier and a power amplifier), the specifications are:

- ☐ Continuous average power output
- ☐ Dynamic headroom
- ☐ Frequency response
- ☐ THD
- ☐ Sensitivity
- ☐ A-weighted signal to noise ratio
- ☐ Maximum input signal
- ☐ Input impedance

The IHF standards also define a number of additional specifications, which are optional. These include:

- ☐ Clipping headroom
- ☐ Output impedance
- ☐ Wideband damping factor

- ☐ Low-frequency damping factor
- ☐ CCIR/ARM signal to noise ratio
- ☐ Tone control response
- ☐ Filter cutoff frequency
- ☐ Filter slope
- ☐ Crosstalk
- ☐ A-weighted crosstalk
- ☐ CCIR/ARM crosstalk
- ☐ SMPTE intermodulation distortion
- ☐ IHF intermodulation distortion
- ☐ Transient-overload recovery time
- ☐ Slew factor
- ☐ Reactive load
- ☐ Capacitive load
- ☐ Separation
- ☐ Difference of frequency response
- ☐ Gain Tracking error
- ☐ Tone control tracking error

Many people are not aware that the specifications of audio equipment can change over time, even without misuse. Electronic component values tend to drift as they age. Changing component values can obviously affect the specifications and performance. Also, specifications of an audio amplifier can be altered by a repair. Whenever a part is replaced, bias and symmetry adjustments should be carefully redone to meet the original specifications of the unit.

DC AMPLIFIERS

If you follow developments within the high-fidelity industry, you may have noticed that in the last few years (since about 1977), a number of manufacturers have been marketing what they call "dc amplifiers" to audiophiles. There is nothing dishonest in this name, but it probably wasn't the best choice—it can be a little misleading. Previously, "dc" has indicated "direct-coupled" when used in connection with audio equipment. (Coupling was covered in Chapter 1.) A modern dc amplifier is a unit that can amplify very low frequency signals, theoretically down to 0 Hz, or dc.

This may well be a case of overkill in an audio amplifier. After all, even the very lowest note on a giant pipe organ is 16 Hz, which is relatively high compared to dc. Sounds as low as 16 Hz are more felt than heard in any event.

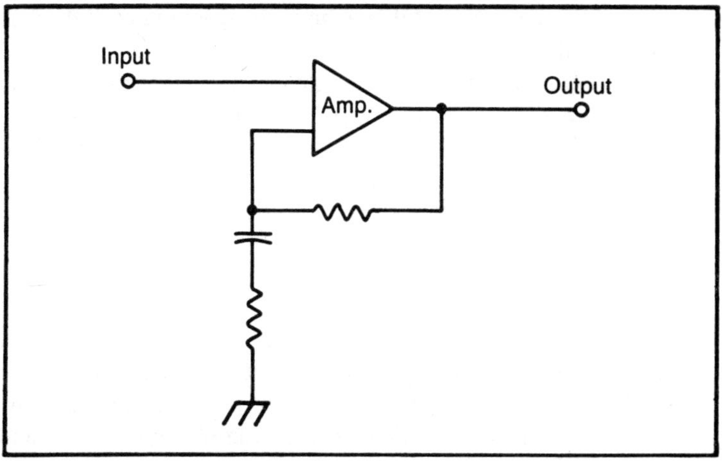

Fig. 5-13. A conventional feedback loop.

To avoid unreasonable goals, several manufacturers define a dc amplifier not as one that can amplify down to 0 Hz, but rather as one with no low-frequency time constants in any of its negative feedback circuits.

Conventional negative feedback loop circuits usually contain reactive components (capacitances and inductances) as illustrated in Fig. 5-13. This can cause frequency sensitive phase shifts, or time delays. Capacitive reactances can be particularly detrimental to low-frequency signals.

A negative feedback loop circuit for a dc amplifier is shown in Fig. 5-14. While the capacitor in the actual feedback network

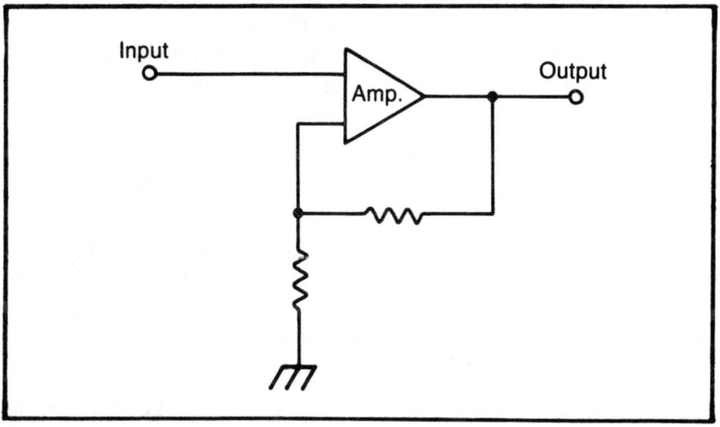

Fig. 5-14. A feedback loop for a dc amplifier.

is eliminated, there may still be coupling capacitors elsewhere in the signal path.

An integrated amplifier (preamplifier, tone control section and power amplifier in a single housing) that has one or all of its sections in a dc configuration may be called a dc amplifier. You should always be sure you know what you are getting.

A dc amplifier is more difficult to design than a traditional amplifier circuit. Why would anyone want to go to the extra bother? For three reasons:

☐ To set up a constant driving power-vs.-frequency relationship between the amplifier and its loudspeakers over the widest possible frequency range. For a dc preamplifier, the same goal applies for the connection between the phono cartridge and the preamplifier/equalizer section.

☐ To eliminate capacitors from all of the signal paths including, but not limited to, the negative feedback loop, as much as possible. Capacitors can cause phase shift and harmonic distortion.

☐ To reduce TIM distortion by improving the waveform transmission characteristics of the amplifier, including the phase and time relationships inherent in complex signals.

The large value electrolytic capacitors used in the output stages are the most likely to degrade sound quality since they can behave as nonlinear impedances even when properly polarized. Such capacitors are likely to affect the coupling between speakers and cartridges and the amplifier. Therefore, these components were soon deleted from modern high fidelity amplifiers.

Even if the output capacitors are eliminated, an integrated amplifier can still have as many as eight capacitors in the signal path, as shown in Fig. 5-15.

Theoretically, it is possible to design an amplifier with no capacitors in the signal path at all. However, this may not work out very well in practical circuits. Occasionally, a signal source, such as a tuner or a tape deck with a slightly leaky output capacitor, may feed a very small amount of direct current to the amplifier along with the signal to be amplified. This undesired dc current will also be amplified by the gain factor of the entire integrated amplifier (often a factor of 30 dB or more). The result is a dc voltage at the amplifier's output terminals, which could damage or destroy an expensive loudspeaker system. To prevent such problems,

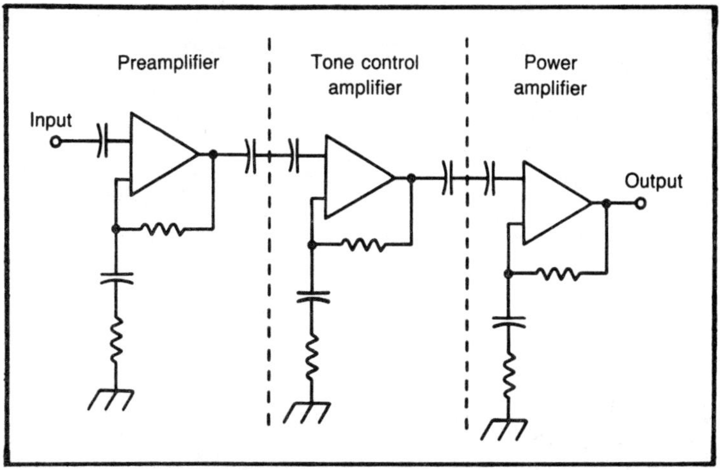

Fig. 5-15. An integrated amplifier without any output capacitors can still have as many as eight capacitors in the feedback path(s).

a few coupling capacitors are almost invariably included in every integrated amplifier design.

In practical integrated amplifiers, some stages are of the capacitorless dc type design, while other stages may be of a more traditional design. Let's glance at a few of the possible combinations.

In the circuit of Fig. 5-16, only the power amplifier uses a pure, capacitor-less dc design. A dc power amplifier stage offers the important advantage of providing excellent damping factors at low frequencies. In audio terms, this gives very well-defined and tight bass reproduction. For optimum overall sound reproduction, this design approach requires a rather hefty power supply.

The circuit of Fig. 5-17 incorporates both a power amplifier and a preamplifier/equalizer that has no input capacitor. This arrangement allows for extremely low-level, low-frequency signals to be applied directly from the phono cartridge into the first preamplifier stage. A coupling capacitor would tend to block any low-frequency signals unless they are quite strong. Generally a FET is used in the input stage because of its voltage-amplifying capability. A phono cartridge effectively appears to the preamplifier as a voltage generator. A FET input also features a high input impedance and does not require any bias voltage. There is some difference in the S/N ratio between a FET and a conventional bipolar transistor phono-input circuit when it is measured under short-circuit input conditions, the FET circuit provides an excellent

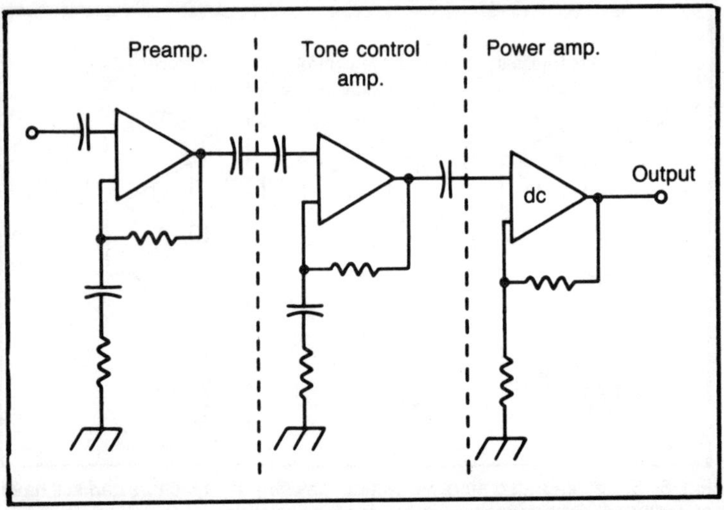

Fig. 5-16. In this integrated amplifier, only the final power amplifier stage has no capacitor.

S/N ratio in actual music listening situations with a phono cartridge connected.

Both the preamplifier and power amplifier stages in Fig. 5-18 use dc design. Also, there is no input capacitor coupling. This arrangement features optimum input coupling for the phono cartridge and optimum output coupling to the loudspeakers.

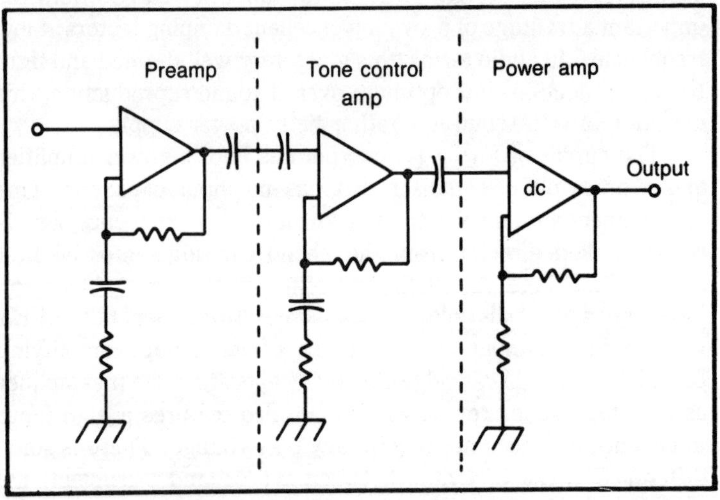

Fig. 5-17. An integrated amplifier with a dc power amplifier stage and no input capacitor at the preamplifier.

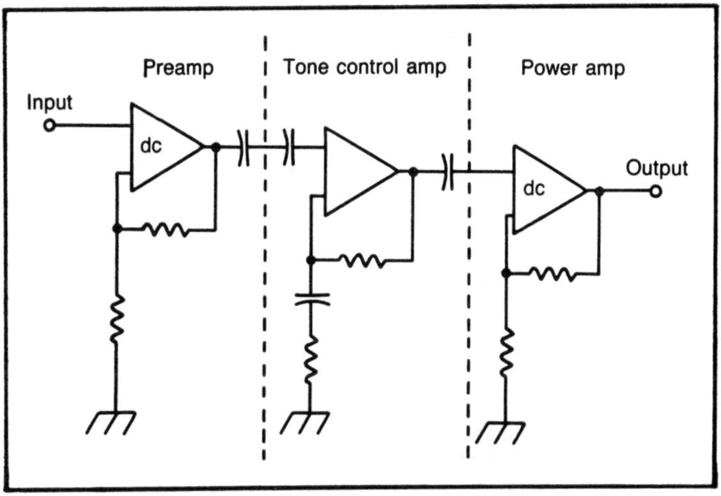

Fig. 5-18. An integrated amplifier with a dc preamplifier and a dc final power amplifier stage.

In Fig. 5-19, capacitors have been eliminated from the feedback loops in both the power amplifier and the tone control amplifier stages. The output capacitor has also been eliminated from the power amplifier stage. Other input and output capacitors are still included between each of the separate sections, forming a low-frequency time-constant circuit.

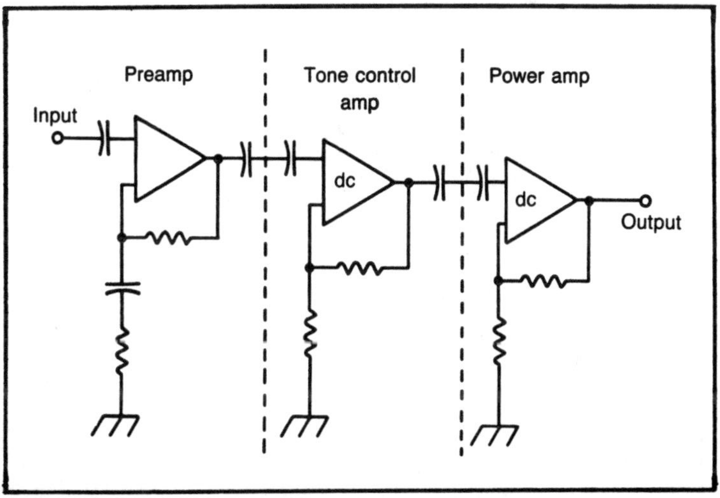

Fig. 5-19. The final power amplifier stage and the tone control section use dc design in this integrated amplifier.

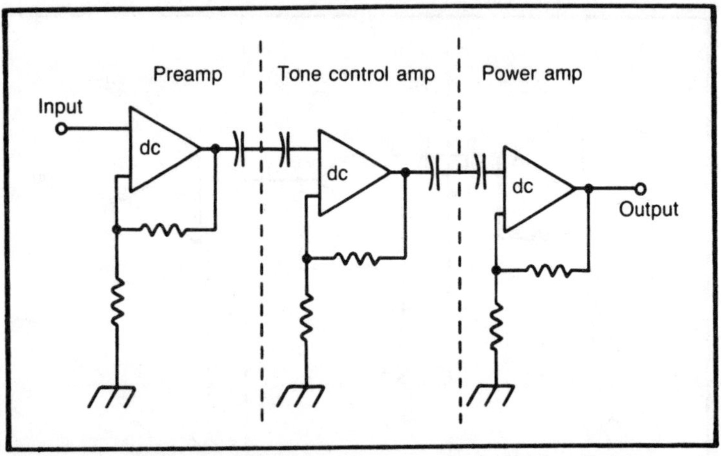

Fig. 5-20. While this integrated amplifier uses dc design for all stages, coupling capacitors are used.

The feedback networks are dc-designed in all of the stages ot Fig. 5-20. Coupling capacitors are still used in the tone control and power amplifier stages.

Finally, Fig. 5-21 is an integrated amplifier in which all of the stages have capacitor-free feedback loops, and all input capacitors to the various sections have also been eliminated. The output capacitors remaining in the circuit (between the equalizer and flat amplifier, and between the flat amplifier and the power amplifier) are designed for a minimum low-frequency time constant.

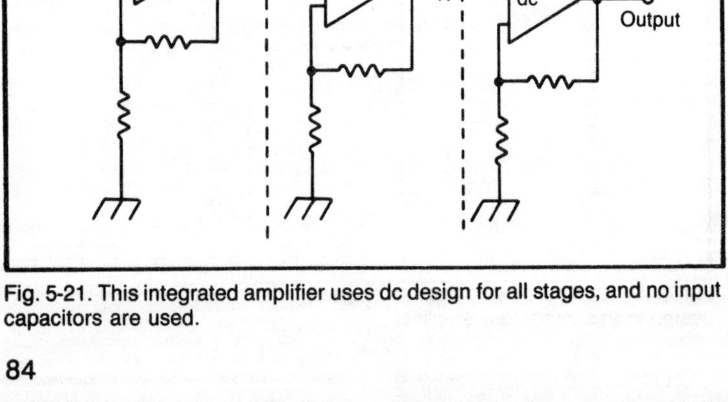

Fig. 5-21. This integrated amplifier uses dc design for all stages, and no input capacitors are used.

SOME PRACTICAL AUDIO AMPLIFIER CIRCUITS

This chapter ends with a few practical audio amplifier circuits that you can build and experiment with. These circuits are fairly simple; a full multi-stage amplifier system is beyond the scope of this book. But these projects, and the preamplifier circuits presented in the next chapter, should give you a good start on designing your own audio system.

I strongly urge you to breadboard these circuits before permanently constructing them. Experiment with various component values—that is the best way to learn. In most cases, the component values listed here are probably the best choices, but by seeing what happens with other values, you can get a better feel of why the listed values are the best. In some cases, you may find a way to improve the operation of one or more of these projects. If so, more power to you!

Headphone Amplifier

The simple one-transistor circuit shown in Fig. 5-22 is ideal for driving headphones. It really isn't quite powerful enough to drive a loudspeaker, although you could try it with a small speaker. An output transformer may be required. The parts list is given in Table 5-1. Nothing is particularly critical. Feel free to experiment with

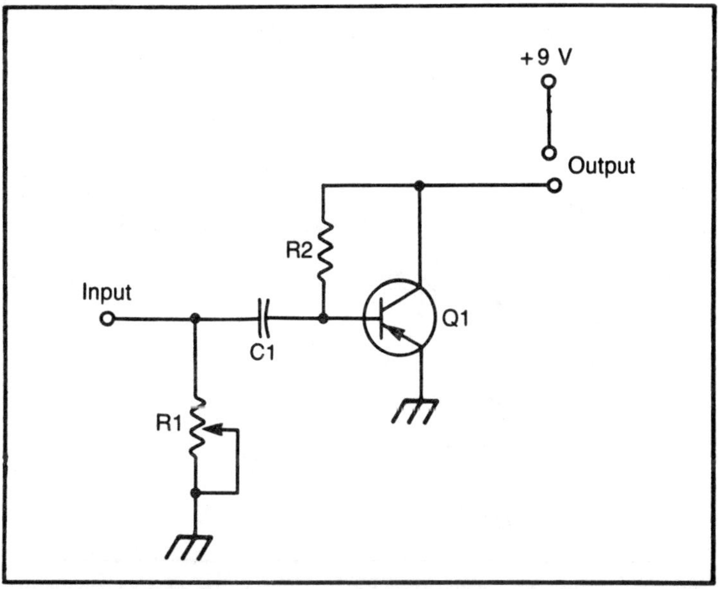

Fig. 5-22. This amplifier circuit is suitable for driving headphones (Project #6).

Table 5-1. Parts List for Project #6.

Component Number	Description
Q1	2N46 transistor (or similar)
C1	0.1 μF capacitor
R1	100 K potentiometer
R2	270 K resistor

substitutions. While the supply voltage is shown here as 9 volts, anything from 6 to 15 volts can be used to power this circuit. The transistor is operated in the Class A mode.

Super-Small Amplifier

In terms of parts count and size, the circuit shown in Fig. 5-23 is probably about as simple as an audio amplifier can get. Besides the single transistor, only an impedance matching transformer, one resistor, one potentiometer, and one capacitor are needed. The potentiometer is used to adjust the bias on the transistor, and may not be needed in all applications. Typical parts values are listed in Table 5-2.

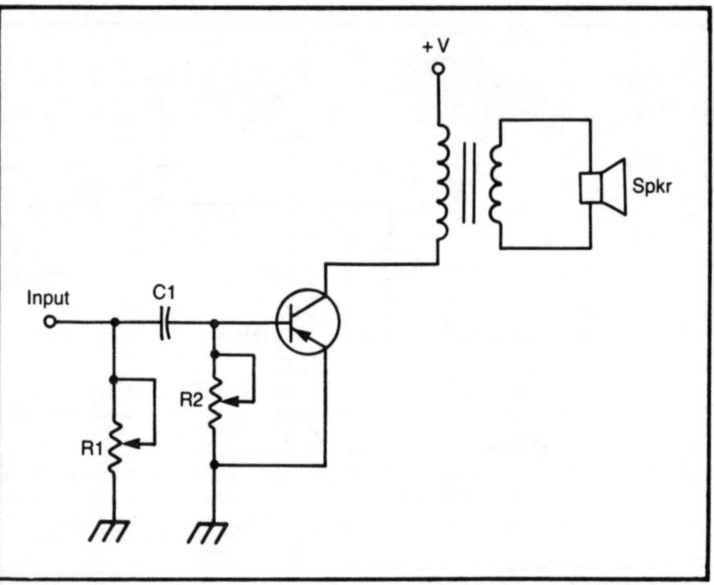

Fig. 5-23. Although very small, this amplifier circuit can produce a healthy volume (Project #7).

Table 5-2. Parts List for Project #7.

Component Number	Description
Q1	2N2956 transistor (or similar)
T1	Output transformer to suit individual application
C1	0.47 μF capacitor
R1	100 K potentiometer
R2	2.5 Meg potentiometer

While small in size, this little circuit can produce a surprising amount of volume through a small loudspeaker.

Ceramic Phono Cartridge Amplifier

The circuit shown in Fig. 5-24 is ideal for amplifying signals from a ceramic-type phono cartridge, or other low voltage signal source. A preamplifier is needed to drive this circuit from a magnetic phono cartridge.

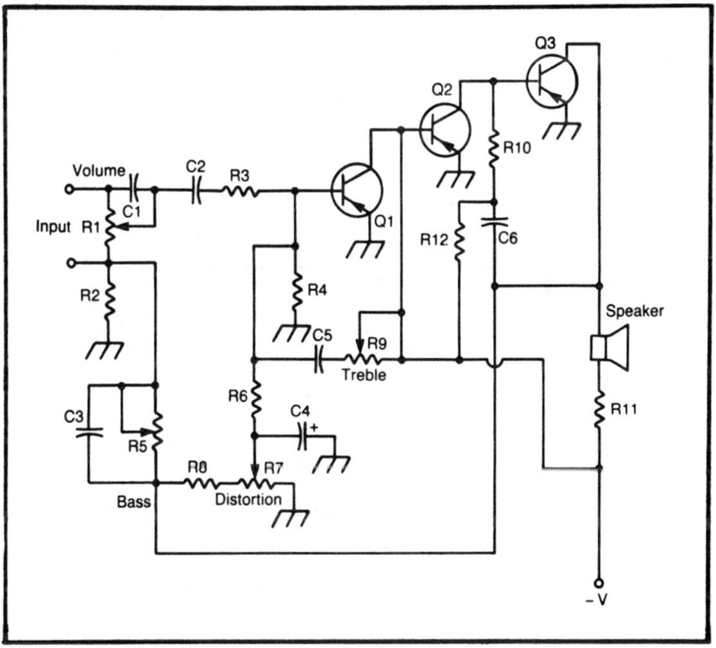

Fig. 5-24. A practical phono amplifier circuit (Project #8).

Table 5-3. Parts List for Project #8.

Component Number	Description
Q1, Q2	SK3004 transistor (or similar)
Q3	HEP200 transistor (or similar)
C1	75 pF capacitor
C2	0.022 μF capacitor
C3	0.1 μF capacitor
C4	15 μF 25 V electrolytic capacitor
C5	0.01 μF capacitor
C6	0.047 μF capacitor
R1	5 Meg potentiometer (volume)
R2	270 ohm resistor
R3	22 K resistor
R4, R8	10 K resistor
R5	15 K potentiometer (bass)
R6	100 K resistor
R7	250 K trimpot (distortion)
R9	100 K potentiometer (treble)
R10	22 K resistor
R11	22 ohm resistor
R12	1.2 K resistor

The parts list for this project is given in Table 5-3. Once again, nothing is terribly critical. Potentiometer R6 should be of the miniature screwdriver adjusted trimpot variety. It should be set

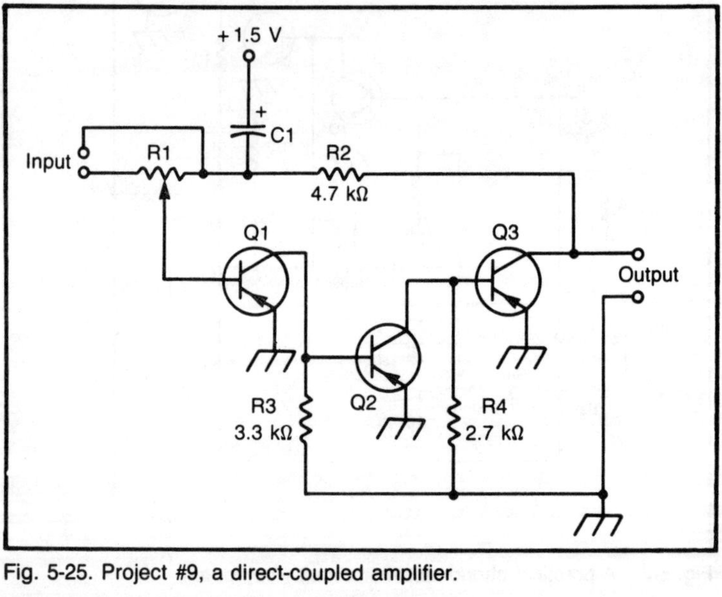

Fig. 5-25. Project #9, a direct-coupled amplifier.

Table 5-4. Parts List for Project #9.

Component Number	Description
Q1, Q2, Q3	2N1382 transistor (or similar)
C1	2.2 μF 10 V electrolytic capacitor
R1	5 K potentiometer (volume)
R2	4.7 K resistor
R3	3.3 K resistor
R4	2.7 K resistor

for minimum distortion, then left alone. Potentiometer R8 is a simple tone control, adjusting the strength of the upper frequencies. If you prefer, a fixed resistor can be used in place of this potentiometer.

The output can drive any small low impedance loudspeaker. While an 8 ohm speaker can be used, better results will be achieved with a 4 ohm speaker. For stereo, you can simply build two of these circuits side by side in a single housing.

Direct Coupled Amplifier

The amplifier circuit of Fig. 5-25 uses direct coupling. Almost any general purpose pnp transistors may be used in this project. All three transistors should be of an identical type.

A parts list for this project is given in Table 5-4.

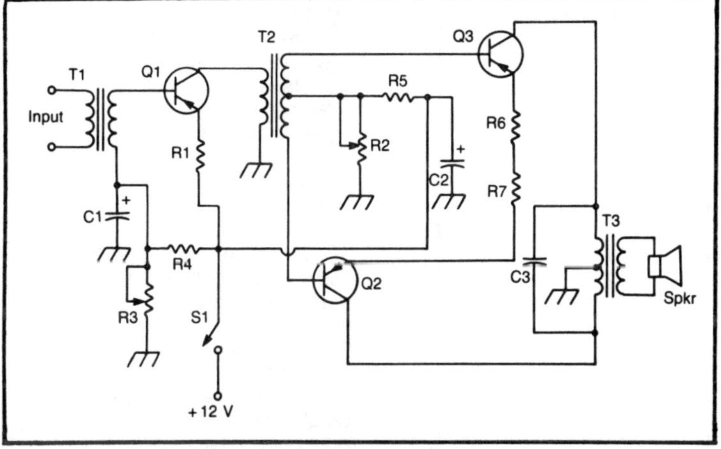

Fig. 5-26. A medium-power amplifier (Project #10).

Medium Power Amplifier

Figure 5-26 shows a simple medium power amplifier circuit. While relatively small and inexpensive, this circuit can put out as much as eight watts.

Potentiometer R5 should be a screwdriver adjust trimpot. It should be adjusted for minimum distortion, then left alone during operation. It would be a good idea to use heat sinks on the transistors, especially Q2 and Q3 which are connected as a push-pull output stage.

A parts list for this project is given in Table 5-5.

Wide Frequency Response Amplifier

The circuit of Fig. 5-27 offers good frequency response over a wide range that extends well beyond the audible range. With the component values listed in Table 5-6, this amplifier passes frequencies from 10 Hz to over 15 MHz. If you use this circuit for rf applications, you may need to use shielding and very high grade capacitors. (See Chapter 7.)

Potentiometer R6 controls the amplifier's gain; in essence, it functions as the volume control.

Table 5-5. Parts List for Project #10.

Component Number	Description
Q1, Q2, Q3	2N1530 transistor (or similar)
T1	Input transformer Primary—to suit input Secondary—100 ohms
T2	Interstage transformer Primary—200 ohms Secondary—300 ohms CT
T3	Output transformer Primary—48 ohms CT Secondary—8 ohms
C1	500 μF electrolytic capacitor
C2	100 μF electrolytic capacitor
C3	0.01 μF capacitor
R1, R5	7 ohm resistor
R2	1 K trimpot
R3	5 K potentiometer (volume)
R4	39 ohm resistor
R6, R7	0.27 ohm resistor

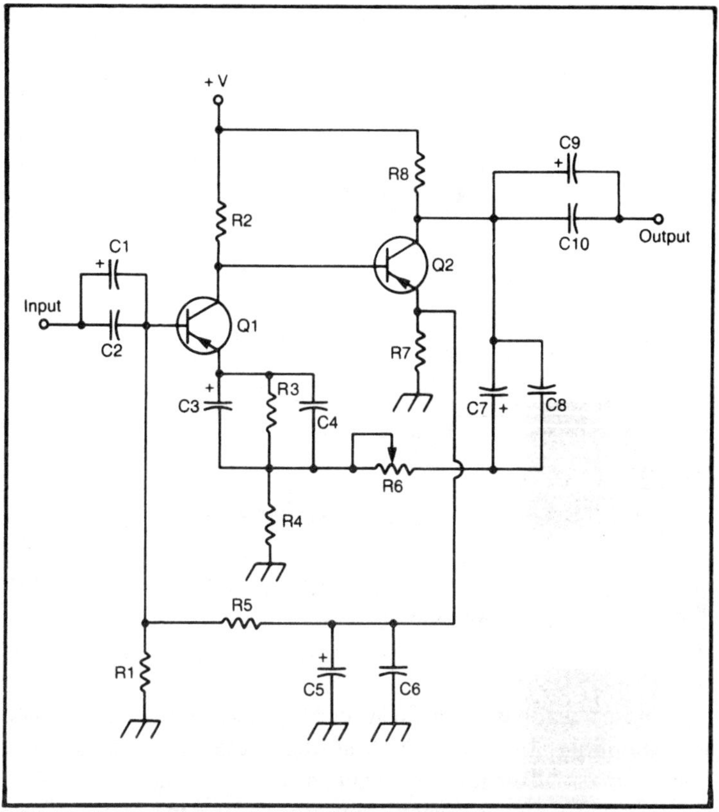

Fig. 5-27. An amplifier with a very wide frequency response (Project #11).

Table 5-6. Parts List for Project #11.

Component Number	Description
Q1, Q2	2N2957 transistor (or similar)
C1, C3, C5, C7, C9	100 μF 25 V electrolytic capacitor
C2, C4, C6, C8, C10	100 pF capacitor
R1	5.6 K resistor
R2	12 K resistor
R3	220 ohm resistor
R4	82 ohm resistor
R5	22 K resistor
R6	10 K potentiometer
R7, R8	1 K resistor

Chapter 6

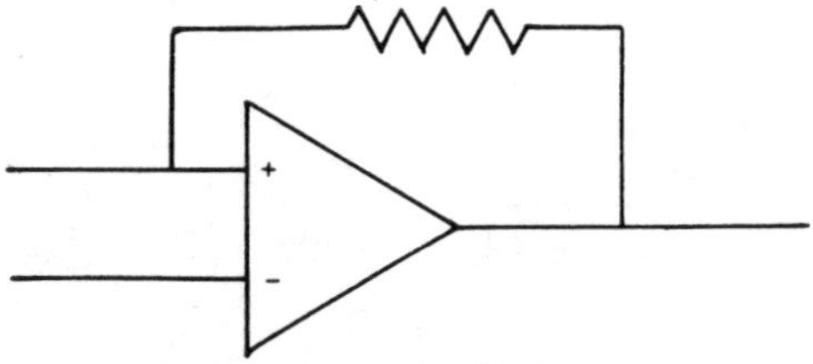

Preamplifiers

IN SOME APPLICATIONS, THE ORIGINAL SIGNAL SOURCE MAY have a very low amplitude. Its level may not be sufficient to drive an ordinary amplifier stage. With very weak signals, any noise is far more critical.

These situations require a special type of amplifier stage called a preamplifier. In many stereo systems, the tone controls or a frequency equalizer may be included in the preamplifier stage.

A number of audio preamplifiers are available in IC form. These include:

- ☐ NE542
- ☐ LM387
- ☐ LM1381

(Amplifier ICs are discussed in Chapter 9.)

BASIC REQUIREMENTS OF A PREAMPLIFIER

There are certain primary requirements for a preamplifier. The necessary gain may range from as low as just 2 or 3 to as much as 5000. The actual gain required for the preamplifier is determined by the input signal required by the power amplifier (or other load) and the amplitude of the original signal source.

For example, if the input signal is 5 mV, but the power amplifier requires an input signal of at least 150 mV, the required gain for the preamplifier is equal to 150/5, or 30. Because the input signal is so small, any noise generated by the preamplifier circuitry is of much more significance than in a power amplifier. In addition, any noise from the preamplifier stage is amplified by all later stages in the system. Consequently, an audio preamplifier must have an exceptionally good S/N ratio.

Audio preamplifiers usually include some sort of frequency compensation in the feedback path. NAB and RIAA compensation (the most common types) are discussed later in this chapter. These compensation networks are the opposite of the pre-emphasis networks used when recording the source material, and are standardized throughout the industry.

PHONO PREAMPS AND RIAA FREQUENCY COMPENSATION

Phono preamplifiers are very common. Inexpensive turntables use ceramic cartridges that put out a fairly high signal level. But ceramic cartridges put a lot of mass on the record grooves. That means that the sound is degraded, and records age rather quickly.

Better-quality turntables use magnetic cartridges, which are much less massive. They respond more quickly to vibrations in the groove walls. Magnetic cartridges also put out very low level signals. A preamplifier is almost always required for a magnetic cartridge turntable.

Phono preamplifiers include a frequency compensation network that complies with the industry RIAA standards. An RIAA compensation network is illustrated in Fig. 6-1. A complete preamplifier circuit with RIAA compensation is shown in Fig. 6-2.

To understand why RIAA compensation is used, let's examine how a record is cut. A chisel-shaped stylus cuts the grooves

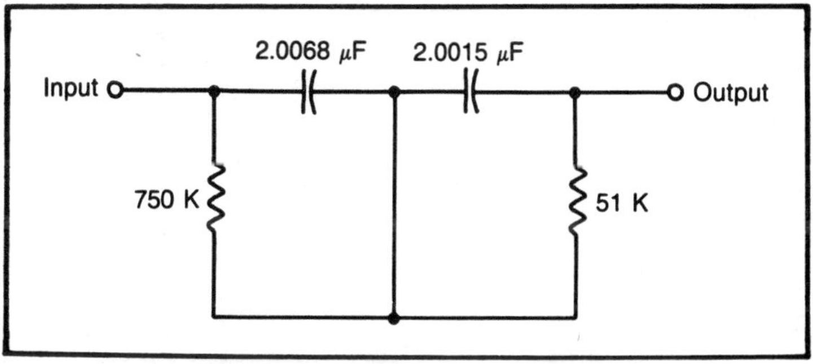

Fig. 6-1. All modern records are recorded with RIAA equalization.

Table 6-1. Parts List for Project #12.

Component Number	Description
IC1	NE542 preamplifier
C1	0.1 μF capacitor
C2	47 μF capacitor
C3	1800 pF capacitor
C4	6800 pF capacitor
R1	47 K resistor
R2	470 ohm resistor
R3	1.5 K resistor
R4	55 K resistor
R5	360 K resistor
R6	1 K resistor

in a stereo disc. The stylus is driven by two vibrating transducers that are mounted at right angles to each other. Modern stereo recordings use what is called a lateral cut. That is, the cutting stylus vibrates from side to side in step with the driving signal. Older monaural records were made with a vertical cut.

As the stylus vibrates, the groove in the vinyl is displaced back and forth about its center. This is called groove modulation. There is a definite practical limit to how far the groove may be modulated. If it is overmodulated, it overlaps with a previous groove on the disc. This condition is known as cutover.

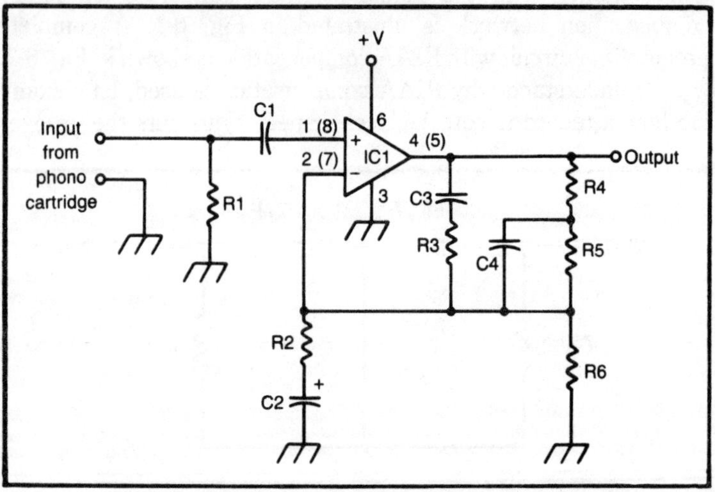

Fig. 6-2. Preamplifier circuit featuring RIAA equalization (Project #12).

Table 6-2. Parts List for Project #13.

Component Number	Description
IC1	NE542 preamplifier
C1	0.1 μF capacitor
C2	120 pF capacitor
C3, C4	2.2 μF capacitor
R1	680 K resistor
R2, R3	220 K resistor
R4, R5, R6	2.2 K resistor
R7	22 K resistor

A trade-off must be achieved between the maximum groove amplitude possible before cutover and the minimum amplitude that provides an acceptable S/N ratio (generally 58 dB). The ratio between these two amplitudes defines the dynamic range of the disc. The dynamic range is typically about 32 to 40 dB.

A record's S/N ratio is governed chiefly by the quality of the vinyl used in the disc. Noise is caused by the grainy characteristic of the disc's surface. High-quality vinyl has smaller grains, and thus, less noise.

To limit surface noise as much as possible, the cutting stylus is heated during the recording process to smooth out some of the graininess. Vinyl surface noise can have a major effect on the final reproduced sound quality. The amplitude of the surface noise can be as much as ten times greater than the circuit noise generated by the preamplifier.

In cutting a record, the audio signal's amplitude is deposited in the form of groove modulation and its frequency is encoded as the rate of change of groove modulation.

For physical reasons, the velocity frequency response of the head (how fast the cutting stylus moves in response to the input signal) has a peak response around 700 Hz, obviously undesirable for any kind of high fidelity application. To correct this, and to produce a velocity output independent of frequency, negative feedback coils are added. When this is done, the cutting head is known as a constant velocity device. The negative feedback pre-equalizes the signal before feeding to the cutting stylus. Low frequencies are attenuated, because they are the most susceptible to cutover. At the same time, high frequencies are boosted to improve the S/N ratio.

The opposite post-equalization must be done in the playback preamplifier to accurately recreate the original recorded signals.

TAPE PREAMPS AND NAB FREQUENCY COMPENSATION

A similar pre-emphasis is used in making recordings on magnetic tape. In this case, the frequency response curve is the NAB standard. The basic NAB compensation network for a playback preamplifier is shown in Fig. 6-3. A complete NAB preamplifier circuit is illustrated in Fig. 6-4. The NAB frequency compensation standard was designed to correct for inherent limitations in the magnetic tape medium.

TONE CONTROLS

In addition to the fixed frequency compensation networks described above, many stereo preamplifiers also feature adjustable tone controls (bass/treble/mid-range). These allow the user some control over the sound produced at the loudspeakers. To a limited extent, these tone controls allow for differences in individual tastes, and to compensate for some inaccuracies of the system's overall frequency response (especially in the speakers and/or the room dimensions).

It is generally convenient to place the tone controls at or near the preamplifier section because the signals are at a low level at this point. Tone controls in the later power amplifier stages would require expensive heavy-duty potentiometers. For the same reason, the system's master volume control is also usually placed in the preamplifier section.

A simple IC preamplifier circuit with tone controls is shown in Fig. 6-5. A typical parts list is given in Table 6-3.

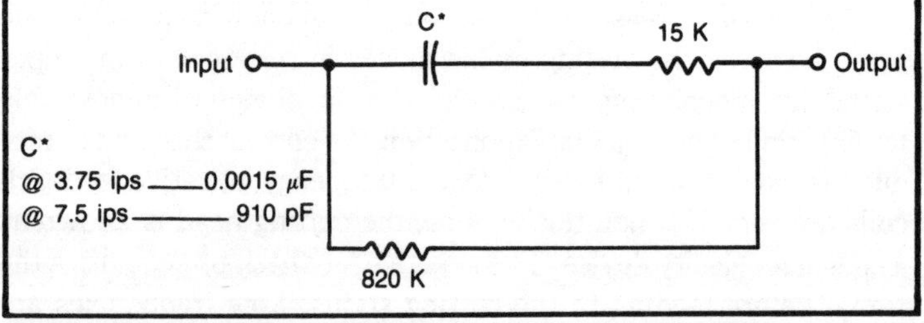

Fig. 6-3. Tape decks usually use a NAB compensation network.

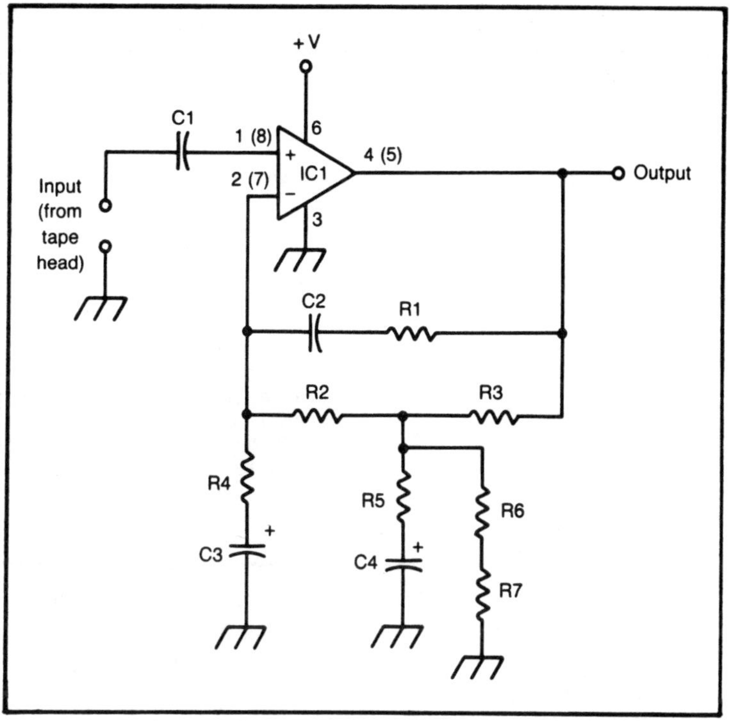

Fig. 6-4. A tape preamplifier featuring NAB equalization (Project #13).

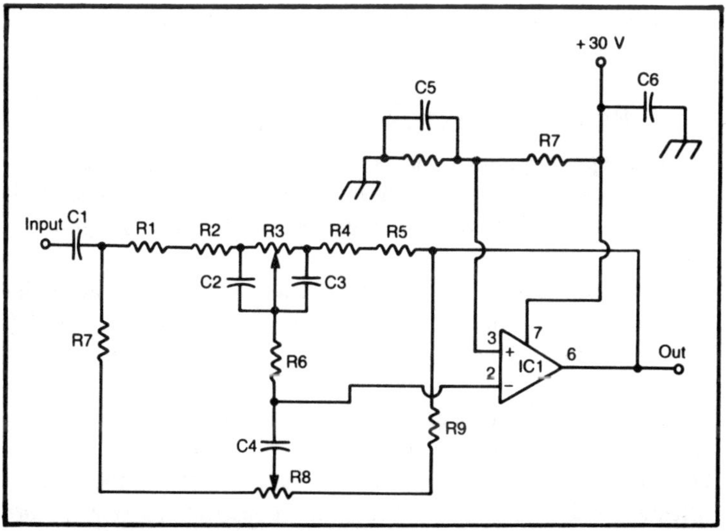

Fig. 6-5. Manually adjustable tone controls are provided in this preamplifier circuit (Project #14).

Table 6-3. Parts List for Project #14.

Component Number	Description
IC1	CA3140 preamplifier
C1	0.047 μF capacitor
C2, C3	750 pF capacitor
C4	22 pF capacitor
C5, C6	0.1 μF capacitor
R1, R4	220 K resistor
R2, R5	18 K resistor
R3	5 Meg potentiometer (bass boost/cut)
R6	2.2 Meg resistor
R7, R9	51 K resistor
R8	5 Meg potentiometer (treble boost/cut)

The tone controls themselves should be linear taper potentiometers. With the component values here, the bass control should give ± 15 dB boost/cut at 100 Hz. The treble control gives ± 15 dB boost/cut at 10 kHz.

When both of the tone controls are set for flat response (no boost or cut), the circuit has unity gain. Additional stages should be used to provide sufficient input levels to later power amplifier stages.

LOW NOISE TRANSISTOR PREAMPLIFIER

In the preamplifier projects in this chapter we have been using preamp ICs, because of their low noise and ease of use. It is generally convenient to use an appropriate IC for a preamplifier circuit.

Sometimes, however, you may want to build a preamplifier circuit from discrete components. The circuit shown in Fig. 6-6 offers low internal noise, and uses just four transistors for both stereo channels.

One rather unique aspect of this circuit is that it has three volume controls. R8 controls the level of the right channel only, R18 controls the left channel only, and R10 is a master volume control, affecting both channels in unison. In most applications, one or more of these potentiometers can be replaced with fixed resistors or trimpots. The choice depends largely on whether or not you need

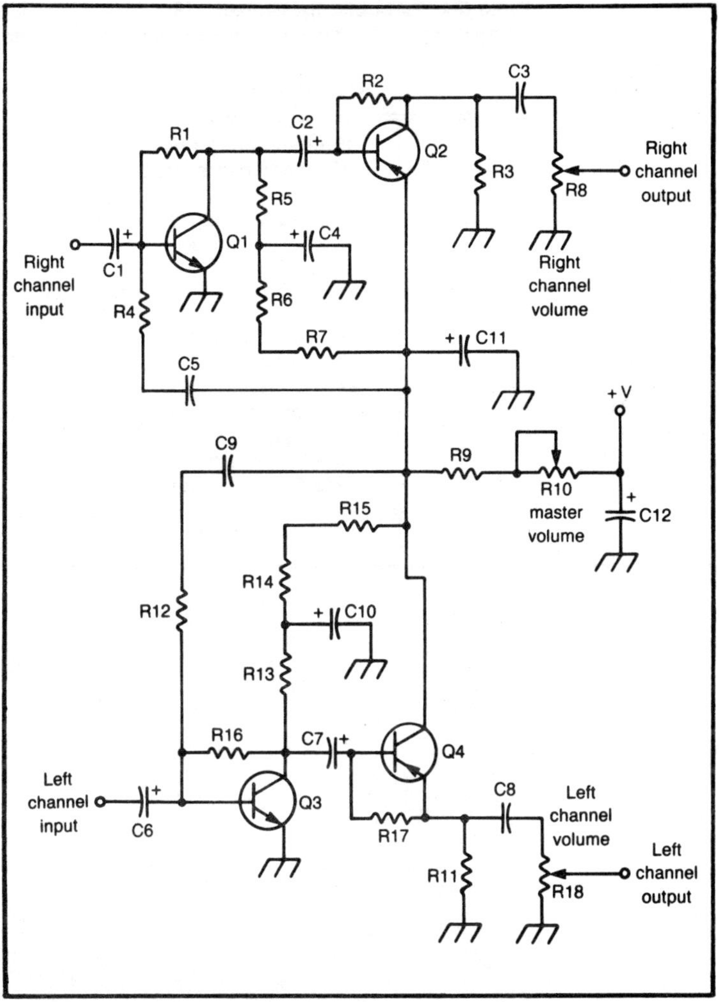

Fig. 6-6. A low-noise transistor preamplifier (Project #15).

to control the stereo channels individually for your specific application.

A complete parts list for this is given in Table 6-4. Nothing is too critical in this circuit, but it should probably be housed in a metalic box to prevent any stray rf pickup. Wires and component leads can act as miniature antennas. Since a preamplifier amplifies such low-level signals, it doesn't take much rf energy to cause problems in the system. Shielded cables between the signal source and the preamplifier are also a good idea.

Table 6-4. Parts List for Project #15.

Component Number	Description
Q1, Q3	GE8 transistor (or similar)
Q2, Q4	HEP254 transistor (or similar)
C1, C2, C4, C6, C7, C10	25 μF 25 V electrolytic capacitor
C3, C8	0.047 μF capacitor
C5, C9	0.022 μF capacitor
C11, C12	100 μF 25 V electrolytic capacitor
R1, R2, R16, R17	560 K resistor
R3, R11	10 K resistor
R4, R12	33 K resistor
R5, R13	15 K resistor
R6, R14	1.5 K resistor
R7, R15	220 ohm resistor
R8, R18	1 Meg potentiometer
R9	100 ohm resistor
R10	1 K potentiometer

Chapter 7

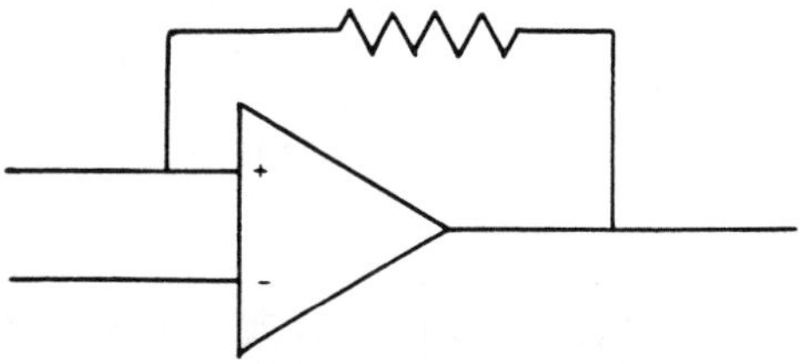

Rf and Video Amplifiers

AMPLIFYING AUDIO SIGNALS (UP TO ABOUT 20 kHz) ISN'T TOO difficult. When you're dealing with higher frequencies, many additional factors come into play. Radio frequency (rf) signals occur at several hundred kilohertz, and well into the megahertz range. Very small stray capacitances can have a significant effect in an rf circuit. In addition, shielding becomes critical at these upper frequencies. Stray rf pickup can occur in almost any short unshielded component lead or connecting wire. This chapter addresses some of the basic problems of rf amplifiers.

CLASS C AMPLIFIERS

Rf amplifiers are often operated in the Class C mode. High power efficiency is usually more important than waveshape fidelity. The rf signal is usually just a sine wave, modulated by the audio signal. Distortion of the sine wave shape will not alter amplitude modulation, frequency modulation, or phase modulation. These aspects of the signal are not affected by the waveshape of the rf carrier. In the receiver, the original sine wave shape can be recreated, if necessary.

As discussed in Chapter 2, in Class C operation, the transistor is biased to conduct for less than one half of each cycle. For an npn device, this is done by making the base negative with respect to the emitter.

Instead of a smoothly varying sine wave, the output is in the form of brief pulses. The transistor is off most of the time. This means that the efficiency can be as high as 85 percent. Very little energy is wasted. For rather obvious reasons of fidelity, the Class C amplifier is virtually useless for any audio application, but this type of circuit is ideal for use in high power rf transmitters.

Typically, a resonant LC circuit is placed at the output. Each time a signal is applied to the amplifier's input, a full cycle is generated across the LC network, if the LC network is tuned for resonance at the frequency (or a multiple of the frequency) of the amplified signal.

CAPACITANCES AND INDUCTANCES IN RF CIRCUITS

Stray capacitances are generally of negligible importance in audio circuits, but in an rf circuit, even the tiniest stray capacitance can have a significant impact upon the operation of the circuit.

A capacitor is simply two conductors (called plates), separated by an insulator (called the dielectric). The capacitance value is determined by several factors, including the size of the plates and the thickness of the dielectric (the separation distance of the plates). From this we can see that a pair of wires can serve as a small capacitor. The air between them can serve as the dielectric. The capacitance is normally extremely small, but capacitive effects become more prominent at higher frequencies. A stray capacitance of a picofarad or so wouldn't cause any noticeable effect at low frequencies, but a signal in the rf region could pass through the "phantom" capacitor and get into part of the circuit where it doesn't belong. Component and lead placement is very critical in rf circuits to avoid the effects of stray capacitances.

A capacitor's reactance (ac resistance) is frequency-sensitive. As the frequency increases, capacitive reactance decreases. The formula for capacitive reactance is:

$$X_C = \frac{1}{2\pi fC} = \frac{1}{6.28\,fC} \quad f = \text{frequency} \quad C = \text{capacitance}$$

An inductor functions in just the opposite way. As the frequency increases, the inductive reactance also increases. The formula for inductive reactance is:

$$X_L = \frac{1}{2\pi fL} = \frac{1}{6.28\,fL} \quad f = \text{frequency} \quad L = \text{inductance}$$

Both inductors and capacitors store energy. That energy is applied to those components when a voltage is placed across their terminals and a current is fed through them. In an ideal capacitor or inductor, all of the stored energy is eventually returned to the circuit, no matter what the operating frequencies may be. Of course, real world components are never ideal. All practical capacitors and inductors have some element of dc resistance associated with them which causes signal losses.

An ideal capacitor would function as a perfectly open circuit to dc signals. Actually, a real-world capacitor functions more like an ideal capacitor and a resistor in parallel. On the other hand, an ideal inductor looks like a short circuit to dc signals. A practical inductor behaves like an ideal inductor and a resistor in series. Because of these built-in resistances, a real capacitor never presents a true open circuit, and a real inductor never presents a true short circuit to dc signals.

It would be handy to know how closely an actual component resembles its idealized form. A specification known as Q, or the quality factor, is used to identify how ideal a capacitor or inductor is. The value of Q compares the amount of energy returned to the circuit and the amount of energy lost in the component due to its inherent resistance. A component that closely complies with the idealized form has a high Q value. In other words, this component suffers less loss than one with a lower Q rating. The reactance and series resistance of an inductor or capacitor is related to the Q value. This relationship is defined by the following equation:

$$Q = \frac{X}{R_S}$$

Similarly, Q is also related to the reactance and parallel resistance by this equation:

$$Q = \frac{R_P}{X}$$

Rearranging the equations, we find that a Q of an inductor is equal to:

$$Q = \frac{6.28FL}{R_S}$$

For a capacitor the Q works out to a value of:

$$Q = 6.28FC\ R_P$$

Sometimes, the quality of a capacitor is described by the dissipation factor, or DF, instead of by Q. DF indicates a loss of energy, usually due to the conversion of that energy to heat. DF is the reciprocal of Q. That is:

$$DF = \frac{1}{Q}$$

The lower the dissipation factor, the better the quality of the capacitor in question.

A resonant network is made up of a capacitor and a coil. Remember, capacitive reactance decreases with increased frequency, while inductive reactance increases with increased frequency. At some specific frequency, the capacitive reactance is exactly equal to the inductive reactance. That is:

$$X_C = Xl$$

This condition is known as resonance. Resonance occurs at only a single frequency for any specific combination of a capacitance and an inductance. The resonant frequency can be found from the component value by using this formula:

$$f_r = \frac{1}{2\pi \sqrt{LC}}$$

There are two possible types of resonant circuits. Either the capacitor and the inductor are wired in series, or in parallel. In a series resonant circuit the impedance is at its minimum value at resonance. Conversely, in a parallel resonant circuit, the impedance's value is at its maximum at resonance.

LIMITATIONS OF ACTIVE COMPONENTS IN HIGH FREQUENCY CIRCUITS

The inherent imperfections of practical inductors, capacitors, and resistors are more serious at high frequencies. An inductor has stray capacitances associated with it. Similarly, a capacitor has stray inductances. Resistors also exhibit some capacitance and inductance. Some types of resistors have so much internal capacitance and/or inductance that they cannot be used in high frequency circuits.

Imperfections of this type are not just limited to passive components. Transistors and other active components also exhibit stray capacitance between the various terminals. A transistor's internal capacitances make up its chief imperfections when used at high frequencies. In some transistors, these internal capacitances are large enough that such devices can only be employed in low-frequency circuits. Other transistors are specially designed for operation up to the high gigahertz range.

Let's consider how bipolar transistors function in rf circuits. For convenience, alpha and beta are normally treated as constant values in low frequency circuits. The only way these values vary from their nominal values is in step along with variations in the collector current. Actually, alpha and beta are not constants. They vary with frequency, becoming smaller in value as the frequency increases. This means that the input impedance of a particular circuit is much higher at low frequencies than at higher frequencies.

Let's look at a typical example. We will use the term $\alpha 0$ to represent the low-frequency current gain between the collector and emitter, that is, low-frequency alpha. Alpha decreases 3 dB to 0.707 of $\alpha 0$ at what is called the alpha cutoff frequency. Various sources represent this frequency with different symbols. The most common of these are:

- ☐ $F\alpha$
- ☐ Fhbo
- ☐ $F\alpha b$

These terms are all interchangeable.

The alpha cutoff frequency can be determined from the transistor's specification sheet. Once this value and $\alpha 0$ are known, the value of alpha at any frequency can be found by using the curve shown in Fig. 7-1.

According to this curve, at F0, alpha is 1/1.4 of its specified low-frequency value. At 2Fo, alpha is 1/2.24 of $\alpha 0$. Increasing the frequency to 4Fo drops the alpha value to 1/4 of its low-frequency figure. Above this frequency, every doubling of frequency drops the alpha value by one half. This relationship holds true on up to extremely high frequencies.

Now the beta decreases at the same rate as the alpha. The beta at any frequency can also be determined using the same curve graph.

It is not difficult to calculate the alpha and beta values at any specific frequency. All you have to do is divide F0 by the frequency under consideration and multiply by the $\alpha 0$ or β value of the transistor involved.

Another important value for rf design that can be found in most transistor spec sheets is the gain-bandwidth product, usually listed as Ft. This is defined as the product of beta and the upper 3 dB limit of a band. This value is useful in determining the beta at any operating frequency. Just imagine that the band stretches from 0 Hz to the desired operating frequency. The beta is derived by dividing Ft by the operating frequency.

At low frequencies, a transistor behaves more or less like a pair of back-to-back diodes, as illustrated in Fig. 7-2. One of the diodes is located between the base and collector terminals. This diode is reverse biased. The second diode is between the base and emitter terminals and is forward biased.

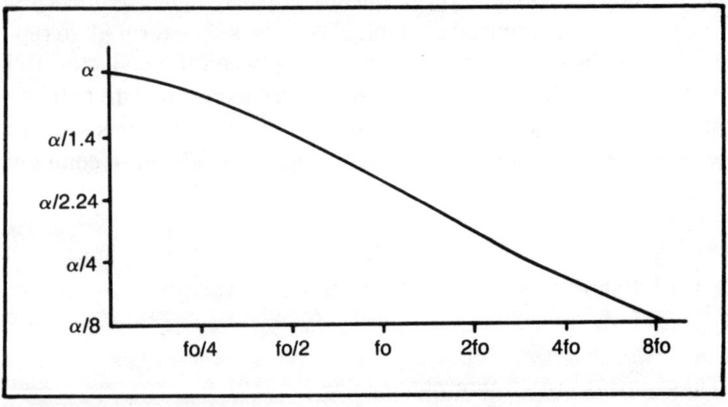

Fig. 7-1. The ac alpha at any frequency can be found from a curve graph like this.

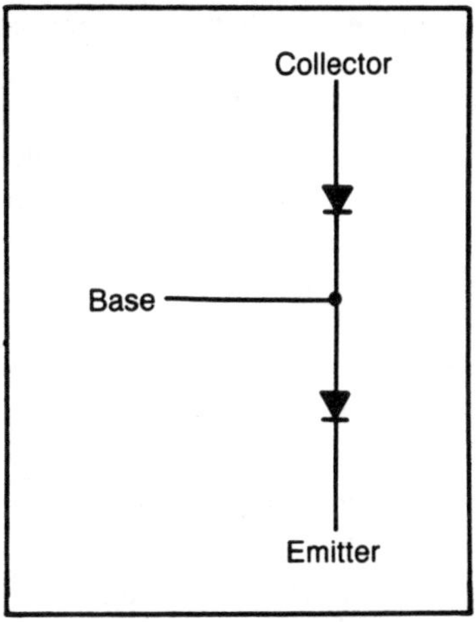

Fig. 7-2. At low frequencies, a transistor behaves like a pair of back-to-back diodes.

You can further simplify this equivalent circuit by replacing the diodes with their internal on and off resistances. The forward biased diode behaves like a low value resistor connected from the base to the emitter (Rba). Similarly, the reverse biased diode could be replaced with a large resistor connected between the base and collector (Rbc).

Transistors also have internal capacitances, so they must be considered as part of the equivalent circuit. At low frequencies, of course, you can reasonably ignore these capacitances, but this doesn't hold true at high frequencies. An equivalent circuit including those capacitances is shown in Fig. 7-3. Notice that almost all of the components in this equivalent circuit are referenced to a point inside the transistor called b′.

Resistance Rbb′ (between the base terminal and point b′) becomes quite significant at high frequencies since it is virtually equal to the high frequency impedance of the various internal capacitances.

Since the b′ reference point is located within the transistor and is inaccessible, it is almost impossible to directly measure the various resistances and capacitances in the equivalent circuit. Fortunately, these values can be estimated using either measurable or specified quantities, including g_m, β, h_{oe}, h_{re}, h_{ie}, and Cob. Except for g_m, all of these factors are included on most complete specification sheets.

g_m is the transconductance of the transistor. This value relates the collector current to the base-emitter voltage. In a FET, g_m relates drain current to gate voltage. Transconductance is approximately the same as the quiescent collector current, in amperes, divided by 0.026.

Another factor is the no-load admittance which is seen when looking back into a transistor. This value is called h_{oe}. Admittance is the inverse of resistance, so the output resistance of the transistor is given by $1/h_{oe}$.

The next factor is h_{re}. To understand this one, let's imagine two voltages, called V1 and V2. V1 is the voltage at the input due to voltage present at the output of the transistor. V2 is the voltage at the output, h_{re} is equal to the ratio V1/V2 (assuming no load at the output).

The impedance seen by looking into the input when the output is short-circuited is called h_{ie}.

Cob is the collector-to-base capacitance of a transistor's common-base equivalent circuit.

All of these parameters (except Cob) are only valid for a common-emitter circuit. If the spec sheets list the h factors for the

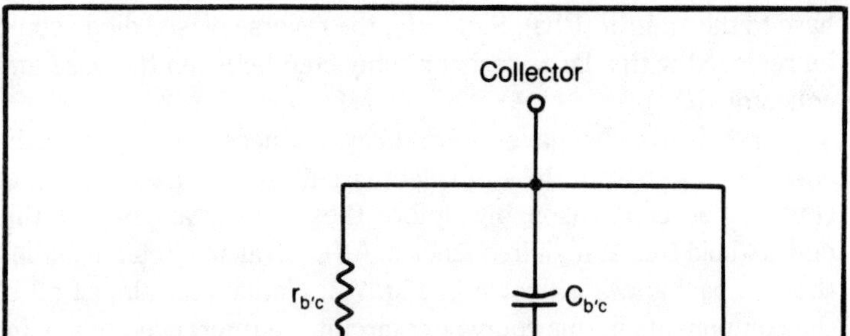

Fig. 7-3. At high frequencies, a transistor's internal capacitances become significant.

common-base circuit (h_{ob}, h_{rb}, and h_{ib}) rather than the h factors for the common-emitter circuit as noted above, use the following conversion equations:

$$h_{oe} = \beta h_{ob}$$

$$h_{re} = \frac{h_{ie}\, h_{oe}}{\beta} + h_{rb}$$

$$h_{ie} = \beta h_{ib}$$

Using the h factors, the values of the resistances and capacitances in the high-frequency equivalent circuit can be found with these formulae:

$$r_{bb'} = h_{ie}$$

$$r_{b'e} = \frac{\beta}{g_m}$$

$$r_{b'c} = \frac{r_{b'e}}{h_{re}}$$

$$r_{ce} = \frac{1}{h_{oe}} - g_m\, h_{re}$$

$$C_{b'e} = \frac{g_m}{2\,\pi\, ft}$$

$$C_{b'c} = C_{ob}$$

At low frequencies, there is no problem in finding the voltage gain. No complex mathematics are required. But voltage gain drops at higher frequencies along with the alpha and the beta. The voltage gain has dropped to 1/1.414 (3 dB) of its low-frequency level at a specific frequency which can be found with this equation:

$$fo = \frac{[\beta + g_m\,(\beta R_e + r_{b'b}| + R_s]}{[2\pi\beta(C_t R_T)]}$$

where:

$$C_T = (C_{b'e} + C_{b'c}\, g_m\, R_L + C_{b'c}\, g_m\, R_e)$$
$$R_T = (R_e + r_{bb'} + R_s)$$

All of the factors in these equations are found as described above, except for Re, RL, and Rs. Re is the value of the emitter resistor. RL is the load resistance and Rs is resistance of the signal source.

Once again, these equations assume a common-emitter circuit. To find the high-frequency voltage gain of a transistor in a common-base or common-collector configuration, follow the usual procedures of low-frequency design, except use the values of α and β, and the effective load on the transistor at the desired frequency in question.

As you can see, amplifier design is much more complex at rf frequencies than for audio frequencies.

I-F AMPLIFIERS

In a radio receiver, there are usually several stages of amplification before the signal is demodulated. The signal picked up by the receiving antenna is quite weak, even when it originated from a powerful broadcasting station.

As we have seen, the higher the frequency is, the more troublesome the design of an amplifier. To limit such high frequency problems as much as possible, the received rf signal is usually down shifted to a lower intermediate frequency (i-f) in the earliest possible stage. The i-f is usually still in the rf region, but low enough to reduce the high frequency amplification problems to a more conveniently handled level.

A standardized intermediate frequency also makes the demodulation process easier. The carrier signal is always at the same frequency (the i-f), no matter what channel is being tuned in.

There are no real differences between an i-f amplifier and a rf amplifier for the same frequency.

SOME PRACTICAL RF AMPLIFIER CIRCUITS

This section takes a very quick look at a few practical rf and i-f amplifier circuits.

Rf Preamplifier

Figure 7-4 shows a rf preamplifier circuit. A parts list for this circuit is given in Table 7-1.

This circuit offers a very high input impedance, minimizing load on the antenna or other preceding stage. The input impedance can be as high as 20 Megohms (20,000,000 ohms).

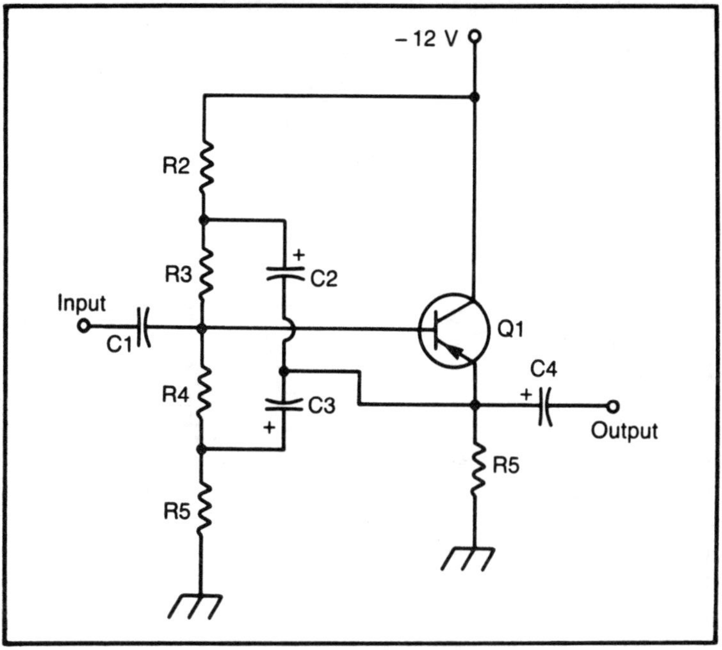

Fig. 7-4. A practical rf preamplifier circuit (Project #16).

This circuit is suitable for long-band reception, up to about 220 kHz. The low end of this preamplifier's frequency response extends below the low end of the audio spectrum (20 Hz).

I-f Amplifier

A typical i-f amplifier circuit is shown in Fig. 7-5. A parts list appears in Table 7-2.

Table 7-1. Parts List for Project #16.

Component Number	Description
Q1	2N2957 transistor (or similar)
C1	0.47 μF capacitor
C2, C3	4.7 μF 25 V electrolytic capacitor
C4	22 μF 25 V electrolytic capacitor
R1	47 K resistor
R2, R3	100 K resistor
R4	56 K resistor
R5	4.7 K resistor

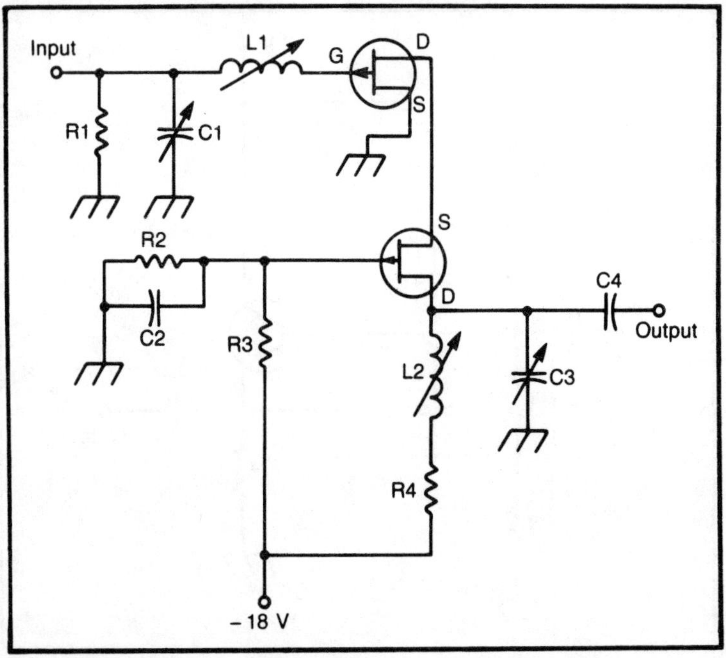

Fig. 7-5. An i-f amplifier (Project #17).

A pair of FETs are cascaded to give a gain of about 20 dB without needing any neutralization, or negative feedback.

This circuit has a very wide bandwidth. With the component values given in the parts list, the bandwidth is 4 MHz, centered around 30 MHz.

Table 7-2. Parts List for Project #17.

Component Number	Description
Q1, Q2	2N4360 transistor (or similar)
C1, C3	365 pF variable capacitor
C2, C4	0.01 μF capacitor
L1	1.5 μH coil
L2	4.5 μH coil
R1	10 K resistor
R2	120 K resistor
R3	470 K resistor
R4	1 K resistor

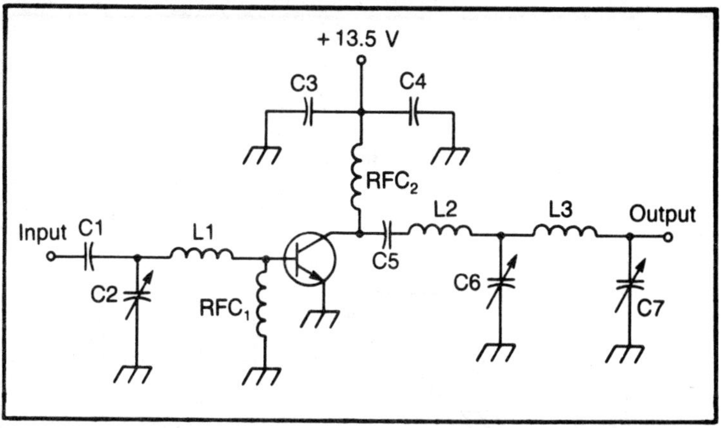

Fig. 7-6. A versatile rf amplifier circuit (Project #18).

Rf Amplifiers

The circuit of Fig. 7-6 is a versatile rf amplifier. A partial parts list is given in Table 7-3. The actual capacitor and inductor values are selected to resonate at the desired frequency in the specific application. For an input signal in the 1 watt range, this circuit puts out about 10 watts. The output impedance is 50 ohms. An alternate rf amplifier circuit is shown in Fig. 7-7. The parts list is given in Table 7-4.

Resistor R1 is optional in many applications. If the circuit is stable (does not oscillate) at the frequency of interest, this component can be eliminated, but if the amplifier is unstable, add the 150 ohm trimpot. To adjust this stability control, set the trimpot to its maximum resistance setting. Slowly reduce the resistance until the circuit becomes stable, use a little paint or nail polish to lock the trimpot's setting, or replace the variable unit with a fixed resistor of the correct value.

Table 7-3. Parts List for Project #18.

Component Number	Description
Q1	2N5646 transistor (or similar)
L1	select for resonance at desired frequency
RFC1, RFC2	rf choke
C1, C5	0.01 μF capacitor
C2, C6, C7	select for resonance at desired frequency
C3, C4	0.001 μF capacitor

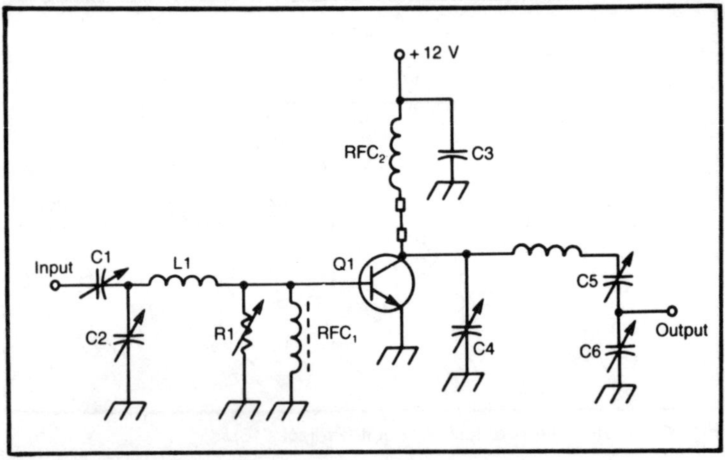

Fig. 7-7. An alternate rf amplifier circuit (Project #19).

VIDEO AMPLIFIERS

Amplifiers used to boost video signals (in equipment such as television sets, VCRs, etc.) are similar to rf amplifiers. The main difference is that a video amplifier requires a much wider and flatter bandwidth. A video signal takes up a lot of space in the rf spectrum. An almost perfectly flat bandwidth of several MHz is essential. Generally, video amplifier design is beyond the scope of the hobbyist.

Table 7-4. Parts List for Project #19.

Component Number	Description
Q1	2N5705 transistor (or similar)
L1, L2	4 turns of #14 wire—core dimensions diameter = 1/4″ length = 5/8″
RFC1, RFC2	rf choke
C1, C2	65 pF variable capacitor
C3	0.022 μF capacitor
C4, C5	45 pF variable capacitor
C6	250 pF variable capacitor
R1	150 ohm trimpot

Chapter 8

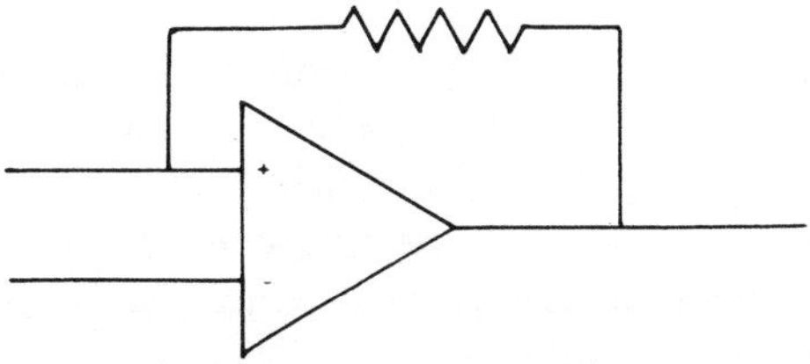

Op Amps

THERE ARE SEVERAL TYPES OF SPECIAL PURPOSE AMPLIFIERS. One of the most important is the operational amplifier, or op amp. This type of circuit has become enormously important in electronics circuitry over the last few years.

In the past, op amp circuits constructed from discrete form were impractical for most applications. They were very expensive and physically unwieldy. The integrated circuit has changed all that. A 741 op amp IC can sit on top of your thumb, and only costs about half a dollar, or less.

Consequently, op amps can be used in many applications today that would have been unthinkably impractical using discrete circuitry. This chapter looks at some of the more popular op amp applications.

WHAT IS AN OP AMP?

The term "op amp" is short for "operational amplifier." This type of circuit was originally designed to perform mathematical operations by analog means. Some of the ancestors of the modern computer were analog rather than digital in nature, and often used operational amplifiers for computation.

Essentially, an op amp is a dc amplifier with very high gain that is controlled by an external feedback network. An op amp has two inputs. One is labelled " – " or "inverting." The inverting input phase shifts the output signal 180 degrees from the input signal.

This also reverses the polarity (positive becomes negative, or vice versa). The other input is labelled "+" or "non-inverting." The non-inverting input does not phase shift the signal, or change the polarity.

Theoretically, an op amp always requires dual power supplies, as shown in the basic schematic symbol (Fig. 8-1). One supply voltage is positive with respect to ground, and the other is negative. Usually the two supply voltages are equal, except for the opposite polarities. The use of a dual supply allows the output signal to swing in either direction from zero (ground potential).

Some modern op amp ICs feature some special circuitry that permits these devices to operate off a single ended power supply, but this involves some electronic cheating, that only confuses the issue on the theoretical level.

THE IDEAL OP AMP

No real-world component is ever ideal, of course, but in studying the theoretical aspects, it is useful to consider the characteristics of an ideal unit.

An ideal op amp features the following characteristics:

- ☐ Infinite open loop gain
- ☐ Infinite CMRR
- ☐ Zero offset
- ☐ Infinite input impedance
- ☐ Zero output impedance
- ☐ Infinite, flat frequency response

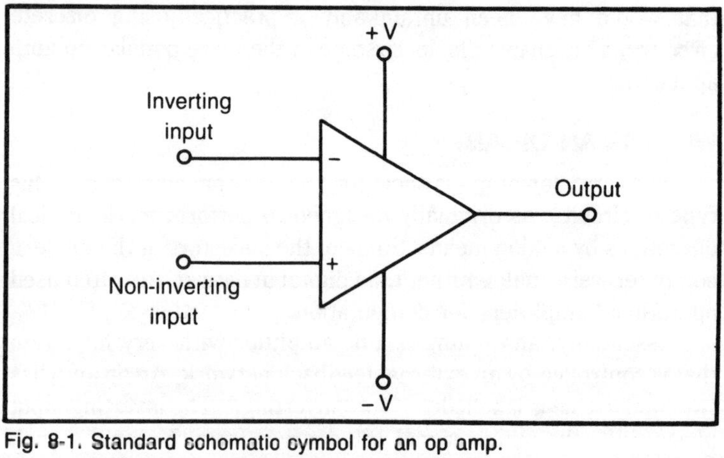

Fig. 8-1. Standard schematic symbol for an op amp.

In most practical op amp circuits, an external feedback loop from the output to the inverting input is used to limit the circuit's gain. This is called a closed loop. Open loop gain is the gain of the circuit with no feedback loop. Ideally, this gain should be infinite. In any practical op amp, of course, there are limitations to the open loop gain. Typical open loop gains for op amp ICs are on the order of 200,000 or so. For most practical applications this is close enough to infinity that the difference can be ignored.

The gain is also limited by the supply voltages. The output voltage cannot exceed the supply voltage. If the op amp is driven past this point, the output signal is clipped. For a dual 15 volt supply, the output voltage can swing over a 30 volt range (from −15 volts to +15 volts).

CMRR stands for Common Mode Rejection Ratio. If the same signal is applied to both the inverting and non-inverting inputs, this is called a common mode condition. Ideally, the equal signals at the opposite inputs should cancel each other out. The output should be zero. This cancellation is called common mode rejection. CMRR is a measure of how effectively the op amp cancels out identical signals at its input. The higher the CMRR figure, the better the cancellation. No practical op amp achieves the ideal of infinite CMRR, of course, but practical op amp ICs normally have respectably high CMRR ratings.

If the inputs are both grounded, the input signals are zero. It follows that the output should be exactly zero also. Unfortunately, minor imperfections in the op amp circuitry can lead to an offset at the output. With zero input, the output has a small but non-zero value (it may be either positive or negative). In practical op amp ICs this offset is very small, but never the ideal of zero.

An ideal op amp has infinite input impedance. It presents no load to its signal source. Practical op amps have a very high, but not infinite, input impedance.

Similarly, an ideal op amp features zero output impedance. The op amp won't be loaded down at all by any later stage. Practical op amps have a very low, but non-zero output impedance.

While not quite infinite, the frequency response of a typical op amp runs from dc (0 Hz) well up into the megahertz region. The response is virtually flat for this entire wide bandwidth.

PRACTICAL OP AMPS

So far we have been dealing with hypothetical ideal op amps for convenience in discussing the basic theory involved. Practical devices, of course, are never perfect, and deviate from the ideal in a number of ways.

The question of just how closely a specific op amp IC approaches the ideal is nominally answered by the manufacturer's spec sheet. The exact information supplied varies according to the manufacturer and the individual device, but most op amp spec sheets contain the following data:

- ☐ A general description of the device
- ☐ A pin-out diagram
- ☐ A schematic, or block diagram of the equivalent internal circuitry
- ☐ Absolute maximum ratings
- ☐ Electrical characteristics
- ☐ Typical performance curves

As an example, consider the typical spec sheet for a type 741 op amp, shown in Fig. 8-2.

For more details on op amp specifications, refer to my earlier book, *How To Design Op Amp Circuits, With Projects And Experiments* (Tab #1765).

THE INVERTING AMPLIFIER

In the majority of op amp applications, only one of the inputs is used. This section examines the basic amplifier circuits using the inverting input.

One of the most common op amp circuits, shown in Fig. 8-3, is known as the inverting follower. The input signal is fed into the inverting input. The non-inverting input is grounded, often through a resistor. The feedback loop limits the infinite gain of the op amp's open loop configuration. In fact, the circuit shown in Fig. 8-4 has a gain of 1, or unity.

Since the inverting input is being used here, the polarity will be reversed at the output. The output voltage is:

$$V_o = -V_iG$$

V_o is the output voltage, V_i is the input voltage, and G is the gain.

Since the gain is 1 in this case, the equation simplifies down to:

$$V_o = -V_i$$

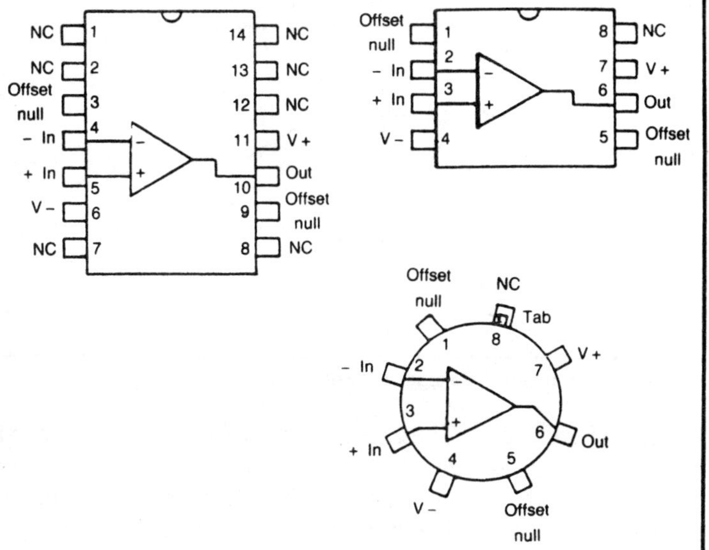

	Minimum	Typical	Maximum
Open-Loop Gain	50,000	200,000	- -
Input Offset Voltage	- -	1.0 mV	5.0 mV
Input Offset Current	- -	20 nA	200 nA
Input Bias Current	- -	80 nA	500 nA
CMRR	70 dB	90 dB	- -
Slew Rate	- -	0.5 V/LS	- -
Input Resistance	1 megohm	6 megohms	- -
Output Resistance	- -	75 ohms	- -
Differential Input Voltage	- -		±30 volts

Fig. 8-2. The 741 is one of the most popular op amp ICs.

The output signal equals the input voltage with the polarity inverted. If the input is 3.5 volts, the output is –3.5 volts. If the input is –2.75 volts, the output is +2.75 volts. In other words, the output is 180 degrees out of phase with the input.

As useful as a unity gain inverting follower can be, you can create an even more useful circuit by adding two resistors, as illustrated in Fig. 8-4. R1 is the input resistor, and R2 is the feedback resistor. The ratio of these two resistances determines

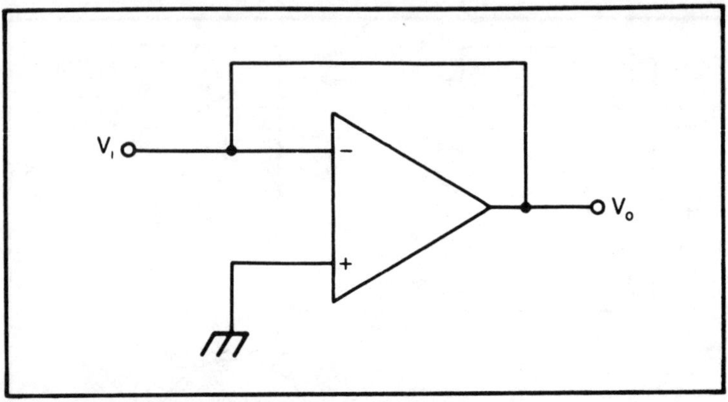

Fig. 8-3. Basic inverting voltage follower circuit.

the gain of the circuit. The gain is negative (less than 1) if the value of R2 is less than R1. If the two resistors have equal values, you once again have unity gain. Finally, (this is the most common arrangement), if the value of R2 is greater than R1, then the gain is positive.

The gain of any combination can be found with this simple formula:

$$G = \frac{-R_2}{R_1}$$

The negative sign indicates the polarity inversion.

The output, remember, is the product of the input and the circuit gain:

$$V_o = V_1 \times G$$

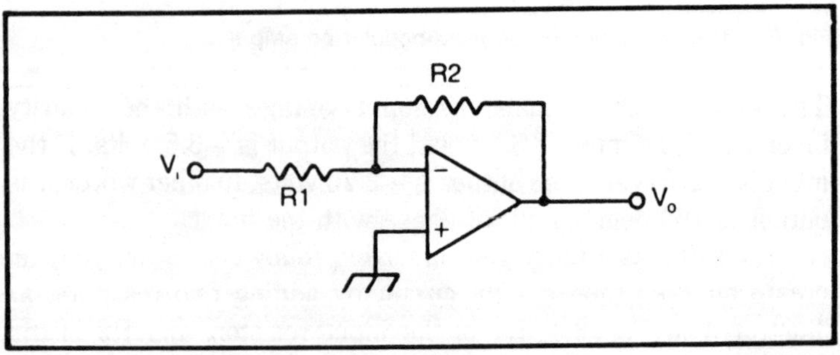

Fig. 8 4. An inverting amplifier with gain.

(Since the non-inverting input is grounded, its input signal is 0, and can be completely ignored in the formula.) These two equations can be combined like this:

$$V_o = V_i \left(\frac{-R_2}{R_1}\right)$$

If this was an ideal op amp, the input impedance would be infinite, placing the input at virtual ground. The significance of this fact is that the resistor values determine the current flow. The input voltage, according to Ohm's law, is:

$$V_i = I_i R_1$$

The output voltage works out in a similar manner:

$$V_o = I_o R_2$$

Again, the negative sign indicates the polarity inversion.

Since an ideal op amp theoretically doesn't draw any current. I_o is equal to I_i. In practical op amps this isn't quite true, but it is a close enough approximation for most purposes. Call the current I, so:

$$V_i = I_i R_1$$
$$V_o = I_o R_2$$

Gain can be defined by the ratio of the input and output voltages. You can make the following substitutions:

$$G = \frac{V_o}{V_i}$$

$$G = \frac{-IR_2}{IR_1}$$

$$G = \frac{-R_2}{R_i}$$

The equal I values cancel each other out. Notice that this gives the same gain equation presented earlier.

Try a few examples. First, assume the following component values:

$$R1 = 1\text{ K}$$
$$R2 = 100\text{ K}$$

The gain works out to:

$$\begin{aligned} G &= -R2/R1 \\ &= -100000/1000 \\ &= -100 \end{aligned}$$

If the input is 1 mV, the output is 100 mV (0.1 volt).

If you change R2 to 47 K, the gain becomes:

$$\begin{aligned} G &= -47000/1000 \\ &= -47 \end{aligned}$$

Changing R1 to 47 K gives us unity gain:

$$\begin{aligned} G &= -47000/47000 \\ &= -1 \end{aligned}$$

Negative gain (attenuation) occurs when R1's value is greater than that of R2. For example:

$$\begin{aligned} R1 &= 100\text{ K} \\ R2 &= 10\text{ K} \\ G &= -10000/100000 \\ &= -0.1 \end{aligned}$$

Very large (or very small) gains are generally not very practical because of the large resistance ratios required. As a rule of thumb, it's best to keep the resistance values between about 1 K and 1 Megohm (1,000,000 ohms).

To get some practical hands-on experience with inverting circuits, carefully breadboard the experimental circuit shown in Fig. 8-5. Virtually any available op amp IC may be used. I have used the pin-outs for the 741 op amp because this device can be found

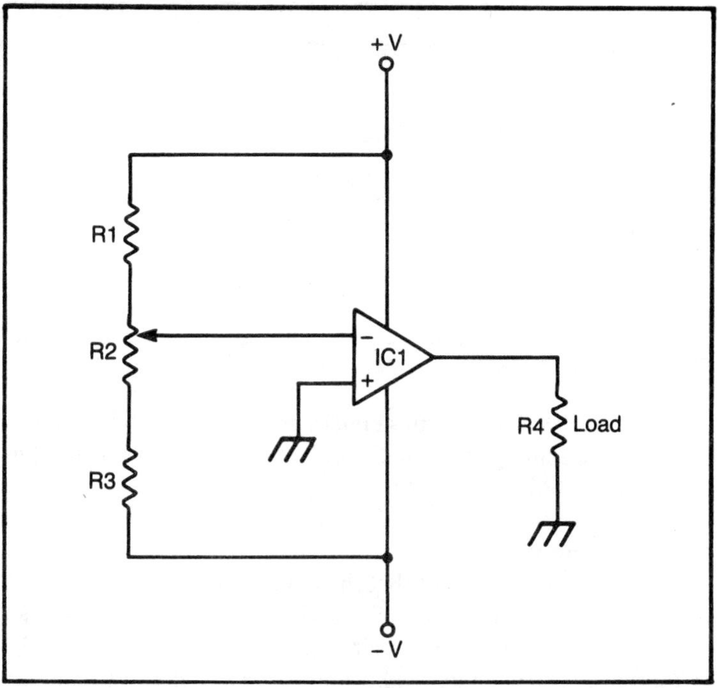

Fig. 8-5. Use this circuit for experimenting with the inverting follower (Project #20).

almost anywhere at very low cost. If you use a different op amp, be sure to check the pin-out diagram to make sure you are connecting the right terminals. For comparison purposes, the pin-out diagram for the 741 in its most common form (8 pin DIP) is shown in Fig. 8-6. Similar substitutions can be made with any of the experiments and projects outlined in this chapter.

The parts list for the inverting voltage follower experiment is given in Table 8-1. You can use a single voltmeter instead of the two shown in the diagram, simply by moving its leads back and forth between points A and B at each stage of the experiment. Point A is the input and point B is the output of the circuit.

For the best results, the voltmeter should be a center-zero type that can measure voltages in either the positive or negative direction.

Resistor R1 and potentiometer R2 form a simple voltage divider to feed different voltages to the inverting input of the op amp. By adjusting the setting of R2, you can determine the input voltage for the circuit.

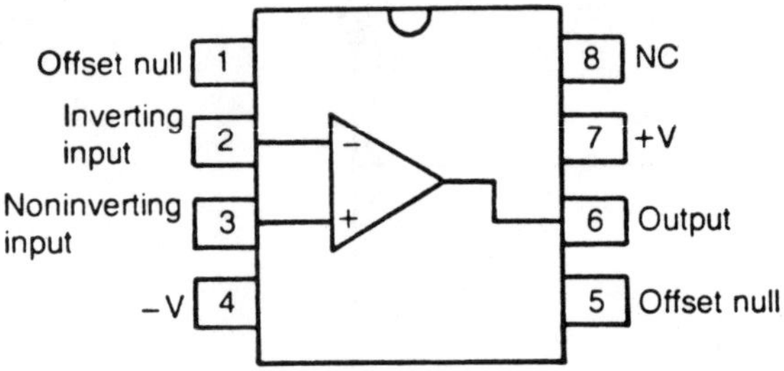

Fig. 8-6. Pin-out diagram for the 741 op amp IC.

Double-check all circuit connections before applying power. This is an extremely good habit to get into. An incorrect connection could conceivably damage or destroy a component (especially an IC), the voltmeter, or even the power supply circuitry of your breadboarding system.

Once you are confident that the circuit is correctly assembled, apply power, and read the voltage at point A (input). Adjust R2 for a reading of exactly one volt. Then read the voltage at point B (output). You should read negative one volt at the output. (Remember this circuit inverts the signal polarity.) There may be a small error due primarily to the chip's internal offsets; however, the output reading should be very close to the input reading, but with the polarity reversed.

Now return to point A and adjust R_2 for an input of 2 volts. Check the output at point B. You should now get a reading of −2 volts. Repeat this procedure for a variety of input voltages.

Table 8-1. Parts List for Project #20.

Component Number	Description
IC1	741 op amp IC
R1, R3	1 K resistor
R2	10 K potentiometer
R4	10 K resistor
	breadboarding socket

If you happen to have an ac signal generator and an oscilloscope handy, you can repeat the same experiment for ac signals. The output signal should always match the input signal, except for a 180 degree phase shift.

For this version of the experiment, remove R_1 and R_2, and feed the ac signal directly into the inverting input of the IC. If you apply very high frequencies (above a few hundred kilohertz or so), the output level may drop off a little indicating that the gain is dropping below unity. This is because of the frequency limitations of the 741. Don't worry about it. This problem should not show up with a high quality op amp IC designed for an extended high frequency range.

The circuit may also break into oscillation at high input frequencies, because there aren't any correction components included in this circuit. Frequency compensation for op amps at high frequencies is discussed later in this chapter. For now, if the output starts looking a little strange at high frequencies, don't worry. Just concentrate on low frequency operation, and you shouldn't run into any problems. This is true for the next few experiments.

After experimenting with the inverting voltage follower, you are ready to explore the inverting amplifier with various gains. A parts list for this group of experiments is given in Table 8-2. The basic experimental set-up is illustrated in Fig. 8-7. Breadboard this

Table 8-2. Parts List for Project #21.

Component Number	Description
Breadboard system	
IC1	741 op amp (8-pin DIP)
R1	1 K 1/2 watt resistor
R2	1 K potentiometer, linear taper
	4.7 K 1/2 watt resistor
	10 K 1/2 watt resistor (2)
	47 K 1/2 watt resistor
Additional resistors, as available	
	100 K potentiometer
R5	4.7 K 1/2 watt resistor
R6	100 ohm 1/2 watt resistor
Hook-up wire	
Voltmeter(s)	

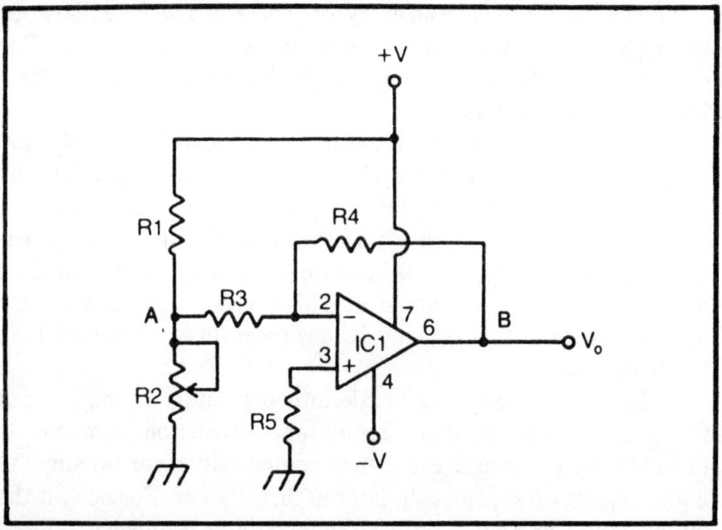

Fig. 8-7. The circuit used in the inverting amplifier experiments (Project #21).

circuit carefully and double-check all connections. Once again, R_1 and R_2 form a voltage divider to provide a variable input voltage.

Remember that the polarity of an inverting amplifier's output is the opposite of the input. If you do not have a center-zero type voltmeter, be careful with the voltage polarities at points A and B.

The gain for this circuit can be found with this equation:

$$G = \frac{-R_4}{R_3}$$

For the first stage of this experiment, use a 1 K resistor for R3, and a 47 K resistor for R4. This gives a circuit gain of about −5:

$$\begin{aligned} G &= -47000/10000 \\ &= -4.7 = -5 \end{aligned}$$

As in the last experiment, set R_2 for an input of one volt at point A. Now measure the output at point B. Remember that the polarity should be inverted at the output. The measured output should be somewhere between about −4.5 and −5 volts. Ideally, it should be −4.7 volts, but resistor tolerances and minor inaccuracies within the IC itself may cause a slightly different output voltage.

Increase the input voltage to 2 volts. Now the output voltage should be about:

$$V_o = V_i \times G$$
$$= 2 \times -4.7$$
$$= -9.4 \text{ volts}$$

A reading between −9 and −9.75 volts is acceptable for the purposes of this experiment. Try calculating and measuring the output voltage for various other input voltages.

For the next phase of this experiment, use the circuit as shown in Fig. 8-8. This version of the circuit feeds negative input voltages into the inverting input of the op amp. Repeat the procedure of the first half of the experiment. You should get similar results, except the output voltages should now be positive.

A final variation of this circuit is shown in Fig. 8-9. Here, potentiometer R_4 can be adjusted to allow for variable gain. Adjust the control while monitoring the output voltage.

These circuits can also be used with ac signals, using the procedure described above.

NON-INVERTING AMPLIFIERS

Of course, in some applications, you don't want the output to

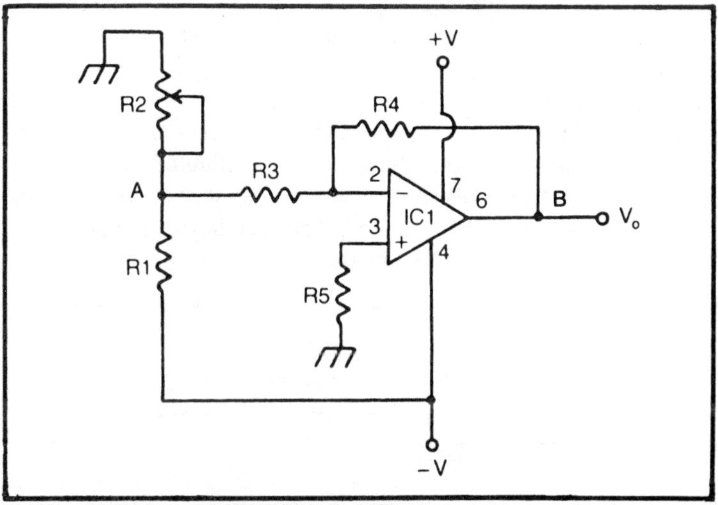

Fig. 8-8. The inverting amplifier circuit can be modified to allow for negative inputs.

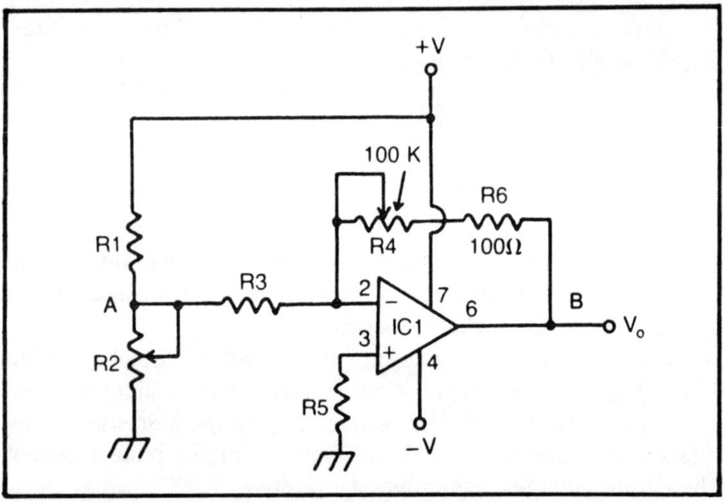

Fig. 8-9. An inverting amplifier circuit with variable gain (Project #22).

be phase shifted, or polarity inverted. In these applications, you must obviously use the non-inverting input.

The basic non-inverting voltage follower circuit is shown in Fig. 8-10. The only difference between this circuit and the inverting version is where the input is fed into the op amp. The feedback loop still connects the output to the inverting input to provide gain limiting negative feedback. The output is in phase with the input and the polarity is not inverted. Since the gain of this circuit is 1 (unity), the output signal is identical to the input signal.

The non-inverting voltage follower is used for isolation and buffering applications. Since the input impedance is extremely high and the output impedance is extremely low, this circuit can also be used for impedance matching. It can be adapted for non-unity gain by placing two resistors into the feedback loop, as in the inverting version (Fig. 8-11). Notice that the resistors are associated with the inverting input, rather than the non-inverting input.

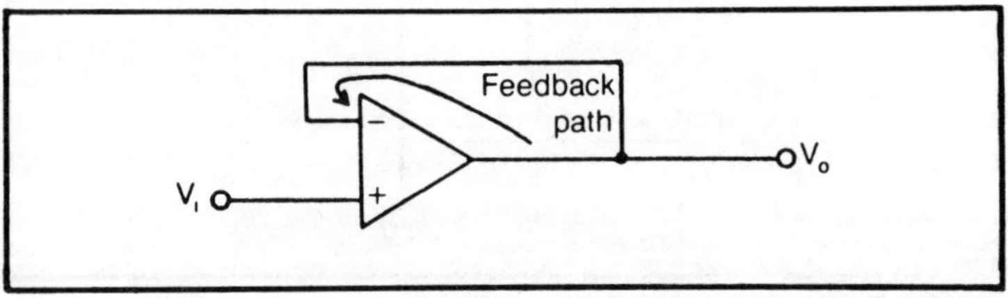

Fig. 8-10. When phase shift is not desired, use the non-inverting voltage follower circuit.

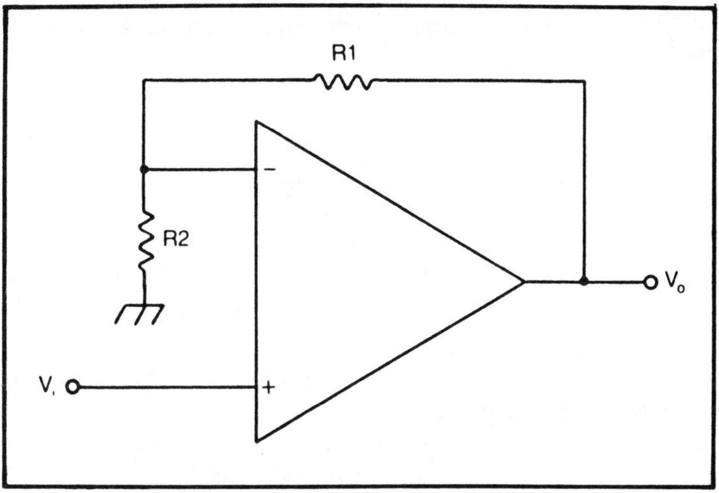

Fig. 8-11. The non-inverting amplifier always has a gain greater than 1.

Once again, the gain is determined by these two resistance values:

$$G = 1 + \frac{R_2}{R_1}$$

No negative sign is used in this equation because the polarity is not inverted.

Since the gain equation adds 1 to the resistance ratio, the gain of this circuit must always be greater than unity, no matter what values are assigned to R1 and R2.

Let's try a few examples. First, make R_1 much larger than R_2:

$$R_1 = 100\ K$$
$$R_2 = 1\ K$$

In this case, the gain works out to:

$$\begin{aligned} G &= 1 + (R2/R1) \\ &= 1 + (1000/100000) \\ &= 1 + (1/100) \\ &= 1.01 \end{aligned}$$

When R_1 is much larger than R_2, the gain is close to unity, but still slightly greater than 1.

Let's reverse the situation, and give R_1 a smaller value than R_2:

$$\begin{aligned} R1 &= 1\ K \\ R2 &= 10\ K \\ G &= 1 + (10000/1000) \\ &= 1 + (10/1) \\ &= 11 \end{aligned}$$

When R_2 is larger than R_1, the circuit gain is significantly greater than unity.

There is one other possible combination. What if both resistors have the same value? Say, 10 K:

$$\begin{aligned} G &= 1 + (10000/10000) \\ &= 1 + 1 \\ &= 2 \end{aligned}$$

When R_1 equals R_2, the gain is 2, no matter what specific resistor values are used.

There is one special case: when both resistors equal 0. (Actually, if R_1 is 0, the value of R_2 is irrelevant.) In other words, there are no resistors at all. The feedback network is just a direct connection from the output to the inverting input. (This is the case in the non-inverting voltage follower circuit of Fig. 8-10). In this instance, the gain works out to:

$$\begin{aligned} G &= 1 + (0/0) \\ &= 1 + 0 \\ &= 1 \end{aligned}$$

When R1 is zero, you have unity gain.

An experimental version of the non-inverting voltage follower circuit is shown in Fig. 8-12. The parts list is given in Table 8-3. Use the same procedure outlined in the inverting voltage follower experiment presented earlier.

The output voltage should always be the same as the input voltage (within a limited margin of error).

Figure 8-13 shows a circuit for experimenting with a non-inverting amplifier circuit. As you should recall from our discussion, the gain of this circuit is always greater than unity. Experiment with various resistor and input values, as in the earlier inverting

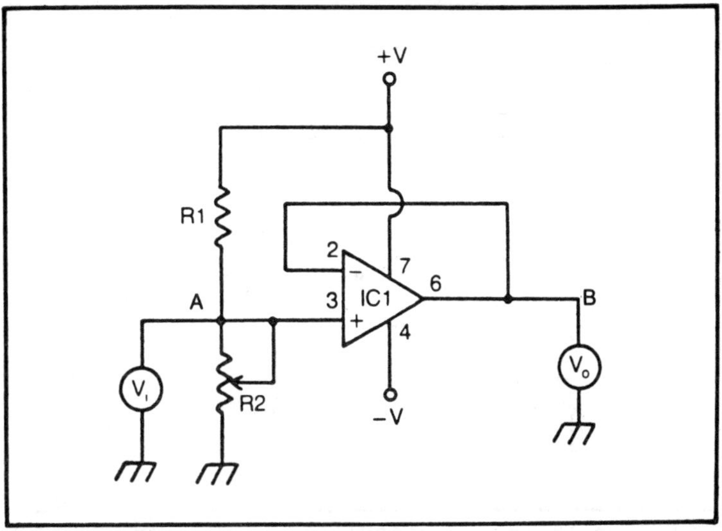

Fig. 8-12. Project #23 involves experimenting with the non-inverting voltage follower.

amplifier experiment. The output polarity should always be the same as the input polarity.

In experimenting with various resistor values, watch that you don't end up with too high a gain. If you feed a high input voltage through a high gain amplifier, the output signal will be clipped. Remember, the output voltage can never exceed the supply voltage.

DIFFERENTIAL AMPLIFIERS

Let's consider what happens when two dc voltages are fed to the inputs of an op amp. Voltage A goes to the inverting input, while

Table 8-3. Parts List for Project #23.

Component Number	Description
Breadboard system	
IC1	741 op amp IC (8-pin DIP—see text)
R1	1 K resistor 1/2 watt
R2	1 K potentiometer, linear taper
Hook-up wire	
Voltmeters (2)	
A single voltmeter may be used—see text.	

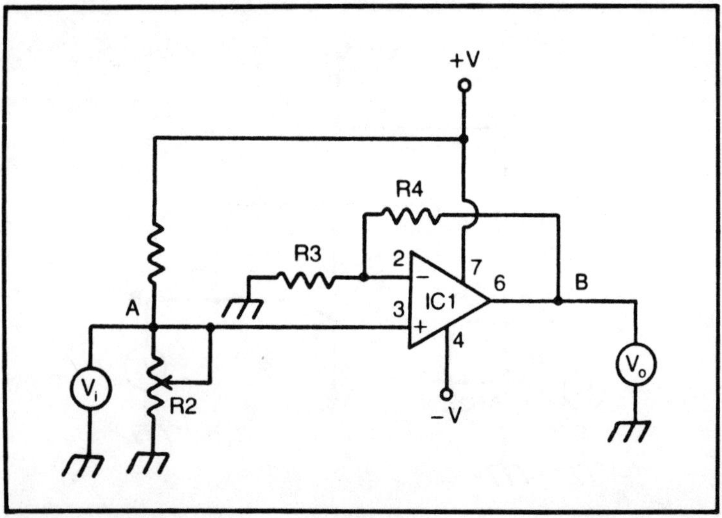

Fig. 8-13. The basic non-inverting amplifier experimental circuit for (Project #24).

voltage B feeds the non-inverting input. There are three possible combinations:

$$A < B$$
$$A = B$$
$$A > B$$

For simplicity, assume that both input voltages are positive. Also, assume that the circuit has unity gain.

If A is less than B, then the output is positive. For example, assuming the following voltages:

$$A = 3 \text{ volts}$$
$$B = 5 \text{ volts}$$

then the output is equal to:

$$\begin{aligned} &(-A) + (+B) \\ &= B - A \\ &= 5 - 3 \\ &= +2 \text{ volts} \end{aligned}$$

If the two input voltages are equal, they cancel each other out, and the output is zero.

If A is greater than B, then the inverting input takes precedence over the non-inverting input. The output (assuming both input voltages are positive) is negative. We can see how this works in an example:

$$
\begin{aligned}
A &= 4 \text{ volts} \\
B &= 2.5 \text{ volts} \\
\text{OUTPUT} &= (-A) + (+B) \\
&= B - A \\
&= 2.5 - 4 \\
&= -1.5 \text{ volts}
\end{aligned}
$$

Of course, gains other than 1 are also possible. The basic operating principles are the same, except the output signal is multiplied by the gain factor.

When an op amp is used in this manner, it is called a differential amplifier. The signal being amplified at the output is the difference between the signals at the two inputs.

A simple circuit for a basic differential amplifier is shown in Fig. 8-14. A feedback loop is always included in a differential amplifier. If you eliminate the feedback and input resistors, you would have a comparator. That type of circuit is discussed in the next section of this chapter.

Differential amplifiers are most frequently encountered in applications in which small signals need to be detected and amplified without being influenced by possibly large interfering signals. This case takes advantage of the CMRR (Common Mode Rejection Ratio) of the op amp.

The interfering common mode signal can come from any of a variety of sources. The most common and troublesome source is electrical pick-up in the connecting leads. This can be a problem especially when fairly long, unshielded wires are used to carry the desired signal from its source to the amplifier.

Under most conditions, if both input wires are of equal length, the interference signal picked up by each wire has an equal amplitude. The same interference signal is applied to both of the op amp's inputs and is rejected as a common mode signal. Only the difference between the two inputs (i.e., the desired signal) reaches the output. This works even if the interfering signal is considerably larger than the desired signal.

Now, refer to the circuit of Fig. 8-14. Generally, R1 is equal to R2, and R3 has the same value as R4. This is not absolutely

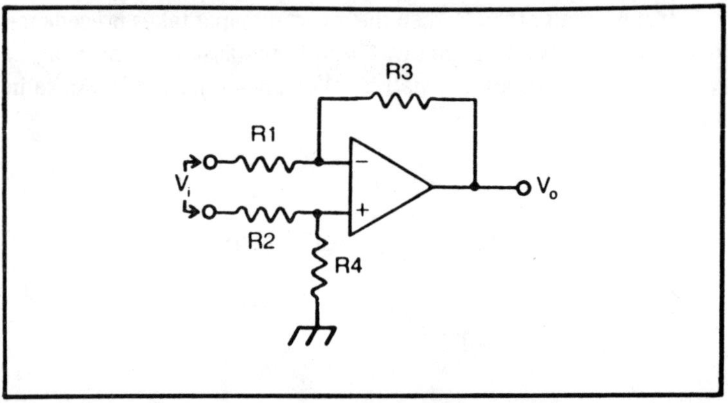

Fig. 8-14. The basic differential amplifier uses both of the op amp's inputs.

necessary, but it is desirable in most practical applications.

Even if different values are used, the R3:R1 and R4:R2 ratios must be equal for the circuit to function properly:

$$\frac{R_3}{R_1} = \frac{R_4}{R_2}$$

These resistance ratios define the closed loop gain of the circuit:

$$G = R3/R1 = R4/R2$$

To see how the differential amplifier circuit works, let's try out a few basic examples. The following examples assume that R1 = R2 and R3 = R4. For the first example, let's go one step further, and assume that all four resistors have identical values, say 22 K. In this case, the gain of the circuit is:

$$G = 22000/22000 = 1$$

If all four resistors have the same value, the circuit exhibits unity gain. This is also the case whenever R1 = R3 and R2 = R4. For instance, assume the following values:

R1 = R3 = 10 K
R2 = R4 = 33 K
G = R3/R1 = 10000/10000 = 1
G – R4/R2 – 33000/33000 – 1

The ratios match. Both are equal to 1, so the circuit maintains unity gain.

For the next example, assume the following component values:

$$R1 = R2 = 33\ K$$
$$R3 = R4 = 100\ K$$

Note that since R1 = R2 and R3 = R4, the resistance ratios are equal by definition. In this example, the gain works out to:

$$G = 100000/33000 = 3.030303$$

or about 3.

If R3 is greater than R1 and R4 is greater than R2, then the gain is more than unity. On the other hand, reversing this relationship by making R3 less than R1 and R4 smaller than R2 results in a gain of less than unity.

As long as the ratios are correct, R1 and R2, and R3 and R4, do not necessarily have to be matched pairs. For example, the following values are perfectly acceptable:

$$R1 = 10\ K$$
$$R2 = 47\ K$$
$$R3 = 100\ K$$
$$R4 = 470\ K$$

Table 8-4. Parts List for Project #24.

Component Number	Description
Breadboard system	
IC1	741 op amp IC (8-pin DIP)
R1	1 K 1/2 watt resistor
R2	1 K potentiometer, linear taper
R3	10 K 1/2 watt resistor (2)
R4	22 K 1/2 watt resistor
	4.7 K 1/2 watt resistor
Additional resistors as available	
Hook-up wire	
Voltmeter(s)	

The relevant ratios match up and present a predictable, consistent gain:

$$G = R3/R1 = 100000/10000 = 10$$
$$G = R4/R2 = 470000/47000 = 10$$

If the resistance ratios do not match up within a reasonable tolerance, the circuit will be unstable and probably will not work.

Since resistor values can deviate from their nominal values, and because of internal imbalances between the op amp's inputs, R4 is often made variable, so that it may be adjusted for the correct ratio match up. A miniature trimpot is generally used for this.

For most applications, the fixed resistors should be 1 percent tolerance types. You may be able to get away with 5 percent tolerance resistors in some non-critical applications, but R4 may be difficult to adjust precisely. It is not advisable to use resistors with tolerances as high as 10 percent or 20 percent.

For high precision applications, especially when dealing with very small signals (on the order of a fraction of a millivolt), it is a good idea to use precision resistors with tolerances of 0.1 percent, or even 0.05 percent. In such applications, R4 should be a high quality ten-turn trimpot.

COMPARATORS

Suppose you eliminate the resistors from a differential amplifier circuit, as shown in Fig. 8-15. In this case you have a comparator. The op amp's full open loop gain is used.

Let's take a look at how a comparator works. For convenience, assume an ideal op amp. There are three possible conditions: the inputs may be equal; input A (inverting) may be greater than input B (non-inverting); or B may be greater than A.

For the time being, assume that all input voltages are positive. If the two inputs are equal voltages, they cancel each other out, and the output is 0, just as in the differential amplifier.

Now, let's increase input A so that it is greater than input B. The output swings negative. But since the open loop gain is so high, even a small difference drives the op amp into clipping. The output almost instantly snaps to its maximum negative value when A becomes greater than B.

If input B is greater than input A, you have just the opposite situation. Now the op amp clips against the positive supply rail. The output snaps to its full positive level.

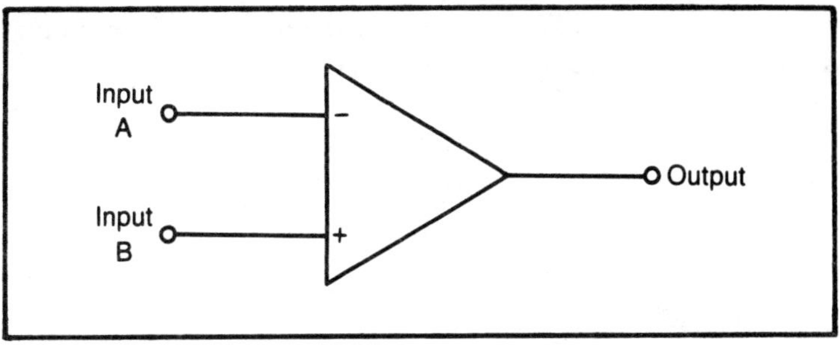

Fig. 8-15. The comparator is similar to the differential amplifier, but the full open-loop gain is used.

Except for an extremely narrow range (completely negligible in most applications), the output can take on one of just three possible values, depending on the input signals:

OUTPUT	INPUTS
0	A = B
−V	A > B
+V	A < B

where V is approximately the op amp's power supply voltage.

The relative values of any two signals can be conveniently compared. (Hence the name comparator.)

Now for some practical experiments on the comparator. The parts list for these experiments is given in Table 8-5.

Table 8-5. Parts List for Project #25.

Component Number	Description
IC1	741 op amp IC
D1, D2	LED
R1, R3, R4, R5	1 K resistor
R2	10 K potentiometer
R6, R7	470 ohm resistor
R8, R9	select for desired reference voltage (see text)
	breadboarding socket

Start out by carefully breadboarding the circuit shown in Fig. 8-16. Use the following resistor values:

R1, R2	22 K
R3	100 K potentiometer
R4, R5	4.7 K

Remember to double-check all connections before applying power to the circuit.

The inverting input of the op amp is fed through R4. This input voltage is determined by the voltage divider made up of R1, R2, and R3. The non-inverting input is referenced to ground (zero volts) through resistor R5. The variable voltage fed to the inverting input is compared to zero (ground potential). Since this is a comparator circuit, there is no feedback loop to limit circuit gain.

The output state of this circuit is indicated by a pair of LEDs. If LED1 is lit, the output is positive. If LED2 is lit, the output is negative. If both LEDs are dark, the output is zero.

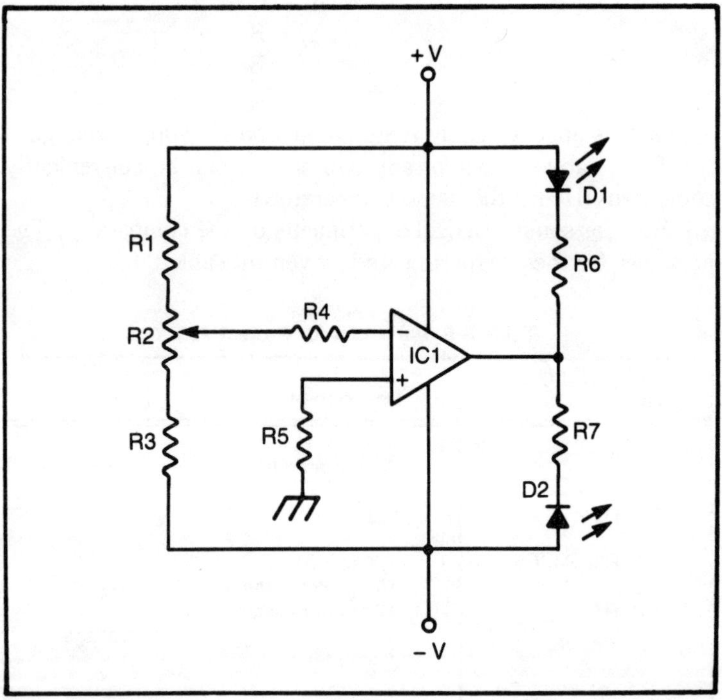

Fig. 8-16. This circuit is used in the first part of the comparator experiment Project 25A.

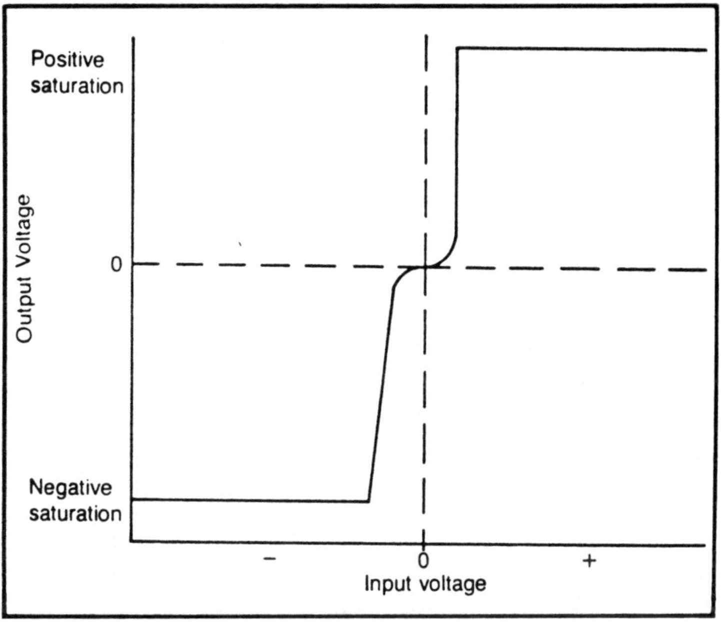

Fig. 8-17. The output voltage graph for the circuit of Fig. 8-16.

Adjust potentiometer R2 for a variety of input voltages, both negative and positive, while observing the LEDs. The LEDs should always indicate the current polarity of the input voltage.

You might notice that for very small non-zero inputs, both LEDs may remain dark, or one may be only dimly lit. There is a narrow crossover band around the zero output point where the op amp is not fully saturated, and clipping does not occur. If you attached a voltmeter to the output and graphed the way the output voltage changes, the resulting diagram would look like Fig. 8-17.

Figure 8-18 shows a simple modification of the basic experimental comparator circuit that permits you to compare the voltage at the inverting input to a non-zero value at the non-inverting input. The relative values of R6 and R7 determine the reference voltage appearing at the non-inverting input.

SUMMING AMPLIFIERS

Operational amplifiers were originally devised to perform mathematical operations. A differential amplifier performs the operation of subtraction. A summing amplifier (sometimes called a mixer) performs the operation of addition.

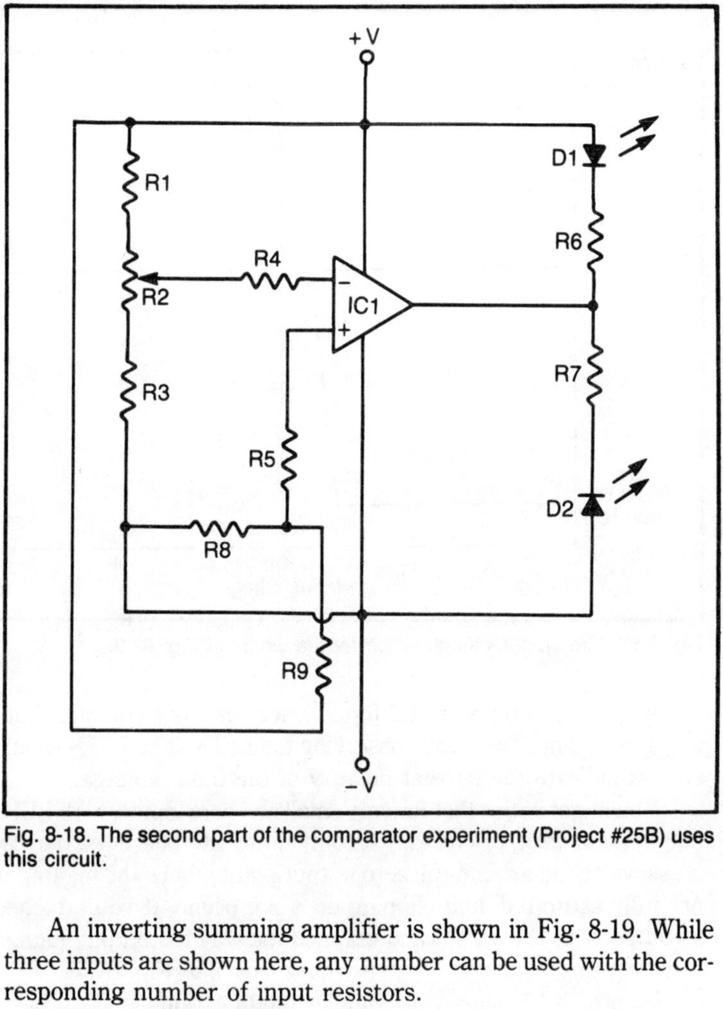

Fig. 8-18. The second part of the comparator experiment (Project #25B) uses this circuit.

An inverting summing amplifier is shown in Fig. 8-19. While three inputs are shown here, any number can be used with the corresponding number of input resistors.

This circuit has all of the features and characteristics of the basic inverting amplifier discussed earlier in this chapter. In addition, the output is equal to the inverted sum of all the input voltages multiplied by their individual gains. For the three-input version shown here:

$$V_o = -\left[V_1\left(\frac{R_4}{R_1}\right) + V_2\left(\frac{R_4}{R_2}\right) + V_3\left(\frac{R_4}{R_3}\right)\right]$$

The negative sign, of course, indicates that the signal is inverted at the output.

If all of the input resistors (R1, R2, and R3) have equal values (R), then the equation can be simplified:

$$V_o = -\left[(V_1 + V_2 + V_3)\left(\frac{R_4}{R}\right)\right]$$

Let's try a few examples. First, assume that all of the gain determining resistors (R1 through R4) have a value of 100 K. Since R1 = R2 = R3, you can use the shorter version of the equation:

$$\begin{aligned} V_o &= -(R4/R) \times (V1 + V2 + V3) \\ &= -(100000/100000) \times (V1 + V2 + V3) \\ &= -(1/1) \times (V1 + V2 + V3) \\ &= -(V1 + V2 + V3) \end{aligned}$$

The various input voltages are simply added together and inverted. All inputs pass through the amplifier with unity gain. For instance, if the inputs are 1 volt apiece:

$$\begin{aligned} V_o &= -(1 + 1 + 1) \\ &= -3 \end{aligned}$$

The input voltages are more likely to take on differing values in most applications. For example:

$$\begin{aligned} V1 &= 1.25 \text{ volt} \\ V2 &= 3.40 \text{ volts} \\ V3 &= 2.35 \text{ volts} \\ Vo &= -(1.25 + 3.40 + 2.35) \\ &= -7.00 \text{ volts} \end{aligned}$$

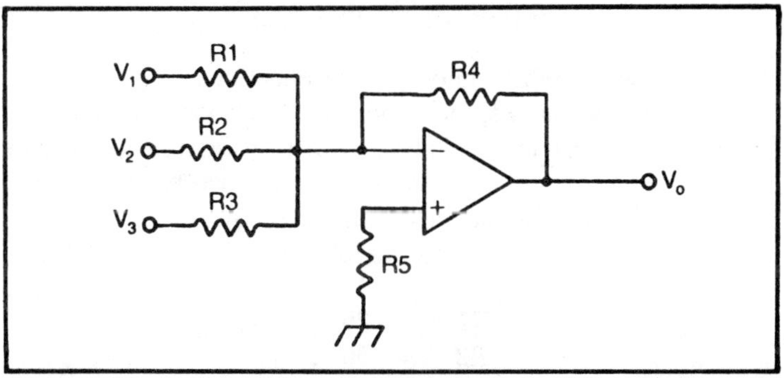

Fig. 8-19. An inverting summing amplifier is used to add multiple inputs for (Project #26).

The input voltages, do not necessarily all have to have the same polarity. For example:

$$
\begin{aligned}
V1 &= 2 \text{ volts} \\
V2 &= -3 \text{ volts} \\
V3 &= 4 \text{ volts} \\
Vo &= (2 + (-3) + 4) \\
&= (2 - 3 + 4) \\
&= -3 \text{ volts}
\end{aligned}
$$

Any combination of input voltages can be summed in this manner.

It is also possible for a summing amplifier to have non-unity gain, of course. This time, assume the following component values:

$$
\begin{aligned}
R1 &= R2 = R3 = 100 \text{ K} = R \\
R4 &= 1 \text{ Meg}
\end{aligned}
$$

Now the general output equation becomes:

$$
\begin{aligned}
V_o &= -\left[(V_1 + V_2 + V_3)\left(\frac{R_4}{R}\right)\right] \\
&= -\left((1000000/100000) \times (V1 + V2 + V3)\right) \\
&= -10 \times (V1 + V2 + V3)
\end{aligned}
$$

The sum is simply multiplied by the gain.

Now let's proceed with the same example, but reduce the value of R4 to 10 K. This time the output signal works out to:

$$
\begin{aligned}
Vo &= -(10000/100000) \times (V1 + V2 + V3) \\
&= -0.1 \times (V1 + V2 + V3)
\end{aligned}
$$

Here we have less than unity gain (attenuation).

In some applications, the various outputs should be weighted differently; that is, they should be amplified by different gains. This is not at all difficult to accomplish, you just have to select appropriate values for the various input resistors.

As a typical example, use the following component values:

$$
\begin{aligned}
R1 &= 25 \text{ K} \\
R2 &= 50 \text{ K} \\
R3 &= 100 \text{ K} \\
R4 &= 100 \text{ K}
\end{aligned}
$$

Notice that the relative values of R1 through R3 follow this pattern:

$$\begin{aligned} R1 &= 1 \\ R2 &= 2 \\ R3 &= 4 \end{aligned}$$

Since the input resistances are not equal, you must use the longer version of the output voltage equation:

$$V_o = -\left[V_1\left(\frac{R_4}{R_1}\right) + V_2\left(\frac{R_4}{R_2}\right) + V_3\left(\frac{R_4}{R_3}\right)\right]$$

$$\begin{aligned} &= -((100000/25000) \times V1) + ((100000/50000) \times V2) \\ &\quad + ((100000/100000) \times V3)) \\ &= -((4 \times V1) + (2 \times V2) + (1 \times V3)) \end{aligned}$$

Input signal V1 will have four times as much influence on the output than input signal V3. V2 will be in between. Each input has its own individual gain. This is called weighting.

To see how this works, let's run through a few quick examples. First assume that all three input voltages are two volts. Solve for the output voltage:

$$\begin{aligned} Vo &= -((4 \times V1) + (2 \times V2) + (1 \times V3)) \\ &= -((4 \times 2) + (2 \times 2) + (1 \times 2)) \\ &= -(8 + 4 + 2) \\ &= -14 \text{ volts} \end{aligned}$$

Naturally, the input signals won't always be of equal value. Assume these inputs:

$$\begin{aligned} V1 &= 1.25 \text{ volt} \\ V2 &= 1.50 \text{ volt} \\ V3 &= 1.75 \text{ volt} \end{aligned}$$

In this example, the output voltage works out to:

$$\begin{aligned} Vo &= -((4 \times 1.25) + (2 \times 1.50) + (1 \times 1.75) \\ &= -(5 + 3 + 1.75) \\ &= -9.75 \text{ volts} \end{aligned}$$

The input polarities may also be mixed:

$$\begin{aligned} V1 &= -2 \text{ volts} \\ V2 &= 3 \text{ volts} \\ V3 &= 2 \text{ volts} \end{aligned}$$

This time the output voltage works out to a value of:

$$\begin{aligned} Vo &= -((4 \times -2) + (2 \times 3) + (1 \times 2)) \\ &= -(-8 + 6 + 2) \\ &= 0 \text{ volts} \end{aligned}$$

Because of its stronger weighting, input V_1 cancels out V_2 and V_3 in this particular case.

In a summing amplifier, the inputs are independent. No interaction takes place between the various input signals. Source V_2, for example, should not load down, or otherwise affect the operation of source V_1 or source V_3.

When using a summing amplifier, watch out for high gains combined with moderate to high input voltages. It is very easy for the sum to exceed the supply voltage limit of the op amp, in which case, the output is clipped.

So far we have ignored resistor R5 in Fig. 8-19. This resistor is included for offset correction. Its value is approximately equal to the parallel combination of all the gain determining resistors:

$$\frac{1}{R_5} = \left(\frac{1}{R_1} + \frac{1}{R_2} + \frac{1}{R_3} + \frac{1}{R_4}\right)$$

In applications requiring maximum precision, R5 can be made variable to permit fine adjustment of the output offset voltage.

FREQUENCY COMPENSATION

Op amp closed-loop circuits are dependent upon negative feedback. Some of the output signal is fed back to the inverting input. At low frequencies, this feedback signal is phase shifted 180 degrees from the original input signal. Some of the input signal is cancelled out, limiting the gain of the circuit.

Unfortunately, as the input frequency increases, so does the phase shifting within the op amp. If the frequency is high enough to produce an additional 180 degrees of phase shift, the difference between the input and output signals becomes 360 degrees. Effectively, they are now in phase with each other. The feedback signal is now added to the original input signal, rather than subtracted. If the closed loop gain is greater than unity, this can lead to all sorts of problems. The op amp can be forced into saturation and clipping. In addition, instability can result, and the op amp could break into oscillation.

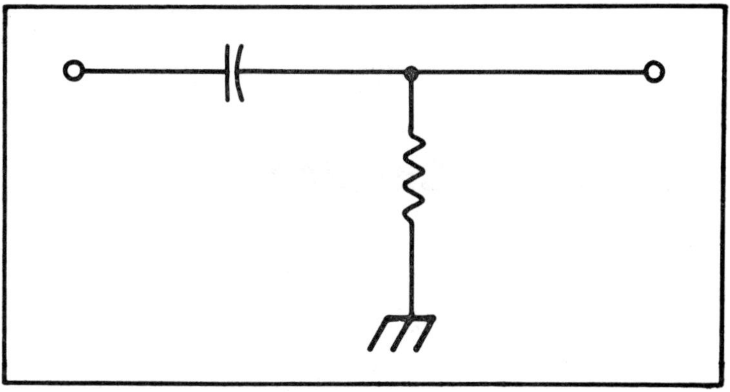

Fig. 8-20. A simple RC network can be added to an op amp circuit for frequency compensation.

Manufacturer's specification sheets usually specify suitable frequency compensation circuits for each device. Frequency compensation is usually accomplished by adding a simple RC network to special output pins included on the IC for just this purpose.

A basic frequency compensating RC network such as the one shown in Fig. 8-20, has a maximum phase shift of 90 degrees. A frequency compensated op amp therefore has a maximum phase shift of 90 + 180, or 270 degrees. This sidesteps the problems of instability and unintentional oscillation that can occur from a 360 degree phase shift of the feedback signal.

Generally, practical frequency compensation is accomplished by adding a small capacitor connected to the specified frequency compensation terminals of the op amp IC. This capacitor combines with the IC's internal resistances to form the necessary RC network.

Many modern op amp ICs include the complete RC frequency compensation network on the chip itself. Op amps with such built-in RC networks are said to be internally compensated. If an external capacitor is employed, the op amp is externally compensated.

Chapter 9

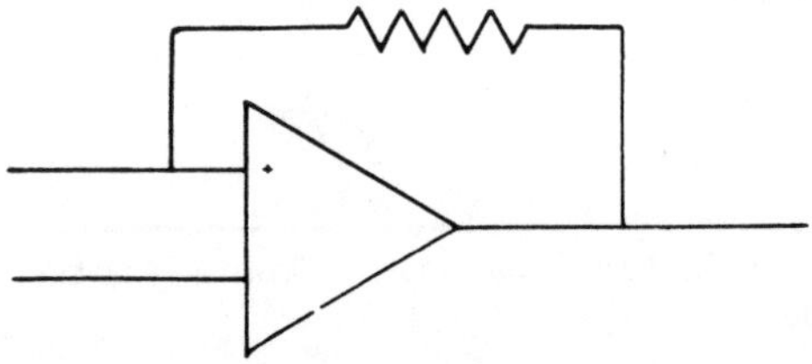

Amplifier ICs

SINCE AMPLIFIERS ARE USED IN ONE FORM OR ANOTHER IN almost every electronic system, it is not surprising that a great many amplifier ICs have been developed for various applications. Some have incredibly impressive specifications. It is often surprising to see how much power can be packed into the tiny IC chip.

For the most efficient and reliable performance, IC amplifiers should always be used with adequate heat sinking. In many cases, the better the heat sinking, the greater the maximum output power can be. When in doubt about how much heat sinking to use, try to err on the side of too much, rather than too little. The chief disadvantage of using a heat sink that is too large is that the device takes up a little more space. The disadvantages of using an insufficient heat sink include inferior performance and possibly thermal damage to expensive components.

It would be futile to attempt to compile a complete list of available amplifier ICs. Hundreds of devices are already on the market, with more appearing every month. A so-called "complete" listing would be out of date before it came off the presses.

This chapter looks at a few representative examples of amplifier ICs of various types. The only kind of amplifier IC that will not be discussed in this chapter is the op amp, because this device was covered in the preceding chapter.

Emphasis in this chapter is on audio amplifier ICs, since these are the most widely available and of the most interest to experimenters.

THE LM380 AUDIO AMPLIFIER IC

The LM380 is an audio amplifier IC that has been around for quite a few years now and still enjoys considerable popularity. Judging from the industrial and hobbyist literature, this chip appears to be the most popular amplifier device around.

The LM380 is widely available, and it comes in two packaging styles, an 8 pin DIP and a 14 pin DIP. The pinout for the 8 pin version is shown in Fig. 9-1, while the 14 pin version is illustrated in Fig. 9-2. Notice that the 14 pin version does not have any additional pin functions. On both packages, only six of the pins are actually active. The remaining pins are shorted to ground, and provide some internal heat sinking. Since the 14 pin LM380 has more heat sink pins than the 8 pin version, it can handle greater amounts of power without overheating. There are no other differences between the two package types.

Without any external heat sinking, the basic LM380 can dissipate up to about 1.25 watts at room temperature. This certainly isn't bad for an amplifier less than the size of your thumbnail, but the LM380 can put out even more power with external heat sinking. For instance, if a 14 pin LM380 is mounted on a PC board with two ounce foil, and the six heat sink pins are soldered to a six square inch copper foil pad, the IC can produce up to about 3.7 watts at

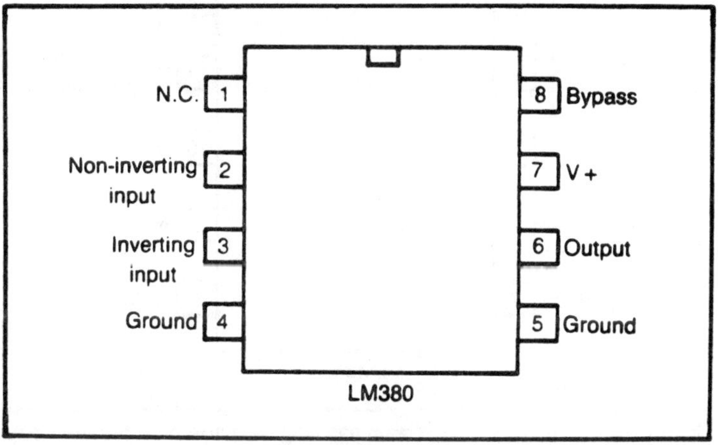

Fig. 9-1. The LM380 amplifier IC, available in an 8-pin DIP housing.

room temperature. This is an impressive almost-threefold increase in power, at very little additional cost.

This chip also features an internal automatic thermal shutdown circuit which turns the amplifier off if excessive current flow causes the IC to start over-heating. This feature significantly reduces the worry of short-circuit problems.

As illustrated in Fig. 9-3, the LM380's internal circuitry is made up of a dozen transistors and associated components. Gain is internally fixed at 50 (34 dB). The output automatically centers itself at one half the supply voltage, effectively eliminating problems of offset drift. If a symmetrical (equal positive and negative voltages) dual-polarity power supply is used with the LM380, the output is centered around ground potential (0 volts), with no dc component to worry about. In this case, no output capacitor is needed to protect the speaker. The LM380 can be wired for the higher fidelity direct-coupled output type of connection described in Chapter 3.

The input stage of the LM380 is rather unusual. The input signal can either be referenced to ground or ac coupled, depending on the particular requirements of the specific application. As

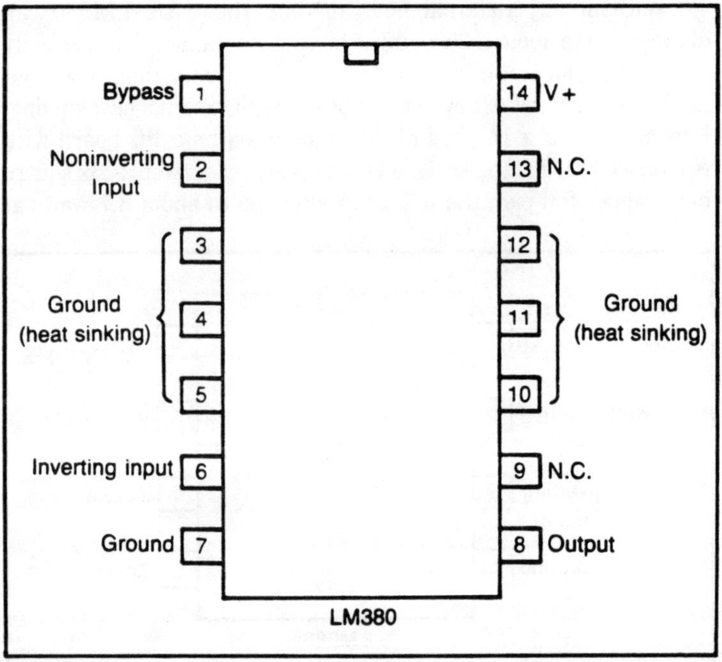

Fig. 9-2. Even the 14 pin DIP version has only six active pins; the extra pins are used for heat sinking.

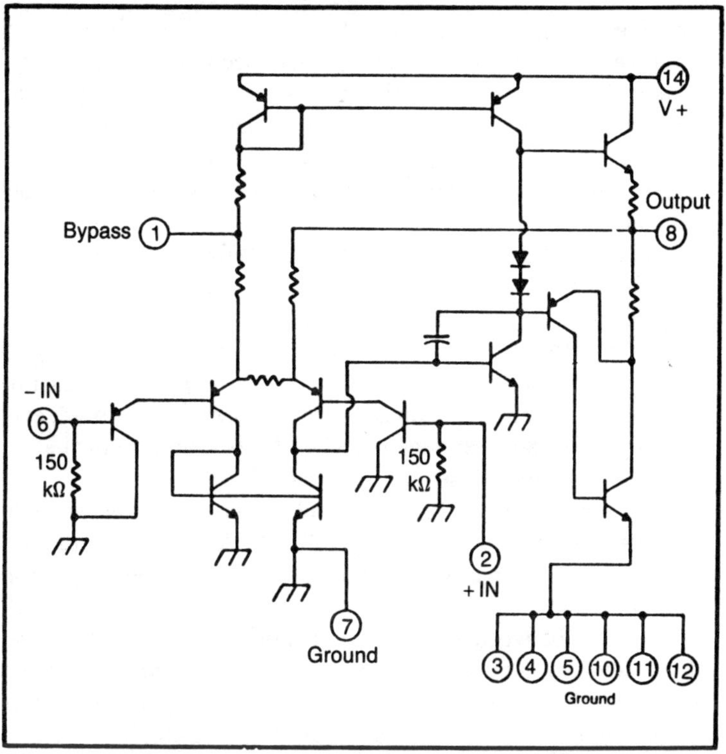

Fig. 9-3. A simplified diagram of the internal circuitry in the LM380 amplifier IC.

you can see, this device offers considerable flexibility to the circuit designer.

The inputs are internally biased with a 150 K on-chip resistance to ground. Transducers or earlier stages which are referenced to ground (no dc component) can be directly coupled to either the inverting or the non-inverting input. (These inputs are similar to those found on op amps. The inverting input phase shifts the signal 180 degrees, while the non-inverting input does not phase shift the signal.)

In most applications, only one of the LM380's inputs is used. There are several possibilities for handling the unused input terminal:

☐ Leave it floating
☐ Short it directly to ground
☐ Reference it to ground through an external resistor or capacitor

In many applications in which the non-inverting input is used, the inverting input is left floating (unconnected). This is fine, but it makes board layout critical. The designer must be on guard for any stray capacitances. While it is always true that stray capacitances may lead to positive feedback, instability, and possible oscillations, this configuration is particularly susceptible to such problems.

The LM380 audio amplifier IC is designed for use with a minimum of external components. The most basic form of a LM380-based amplifier circuit is shown in Fig. 9-4. Clearly, it would be difficult for a circuit to be much simpler than this. The only required external component is the output decoupling capacitor. As mentioned earlier, if a dual-polarity power supply is used, even this capacitor can be eliminated. The LM380 can certainly be considered complete in itself.

Of course, in many practical applications it may be desirable, or even necessary, to add several external components. For example, if the chip is located more than two or three inches from the power supply's filter capacitor, a decoupling capacitor should be mounted between the V+ terminal of the LM380 and ground. Typically, the value of this capacitor is in the neighborhood of 0.1 μF, or so. For best results, this decoupling capacitor should be mounted as close to the LM380's body as possible.

The LM380 tends to become unstable and break into oscillations if it is used in a high-frequency (several megahertz, or

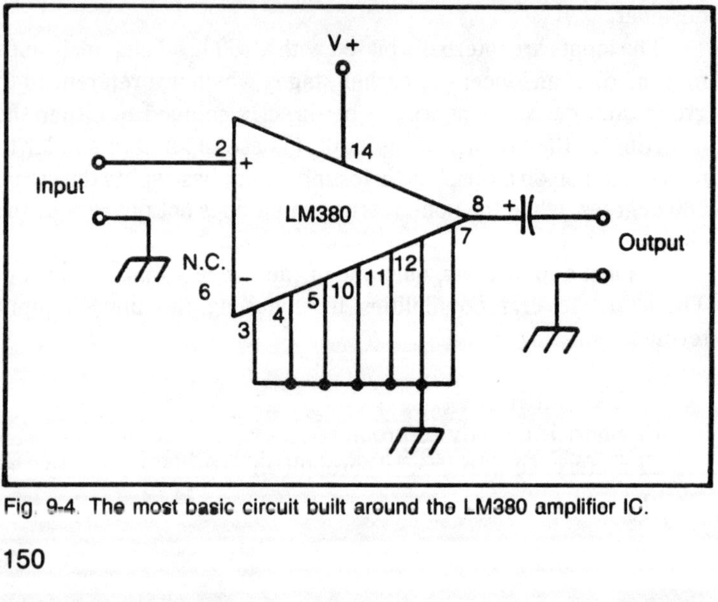

Fig. 9-4. The most basic circuit built around the LM380 amplifier IC.

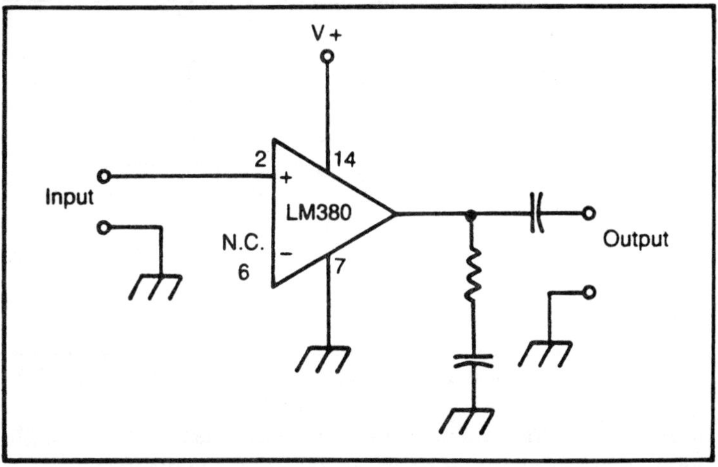

Fig. 9-5. Adding a resistor and a capacitor to the circuit of Fig. 9-4 helps minimize oscillation problems.

more) rather than audio application. Even though this IC is designated as an audio amplifier, it can function as a rf amplifier too. Adding an extra resistor and capacitor, as shown in Fig. 9-5, helps suppress parasitic oscillations in high-frequency applications. Generally, the resistor's value is very small. A typical value is 2.7 ohms. The capacitor is usually about 0.1 μF.

Since these parasitic oscillations only occur at 5 to 10 MHz, they obviously won't be of much importance in most audio applications. Even so, if the LM380 is being used in a rf sensitive environment, such oscillations could pose a problem unless they are properly suppressed.

(Note that in Fig. 9-5 and all the future diagrams, the heat sinking pins are not shown. This is for clarity in the circuit diagrams. The grounding of these pins is assumed in all cases.)

Figure 9-6 shows a practical audio amplifier circuit using the LM380. A typical parts list for this project is given to Table 9-1.

The input to this circuit can be provided by an inexpensive low-impedance microphone, like those sold for use with portable cassette recorders. If a low-impedance source is used, an impedance matching transformer is necessary. If this circuit is to be used with a high-impedance source, this transformer can be eliminated.

The 1 Meg potentiometer serves as a volume control. The fixed gain of this circuit can be increased by adding a little positive feedback.

An 8 ohm speaker can be driven directly by the LM380 audio

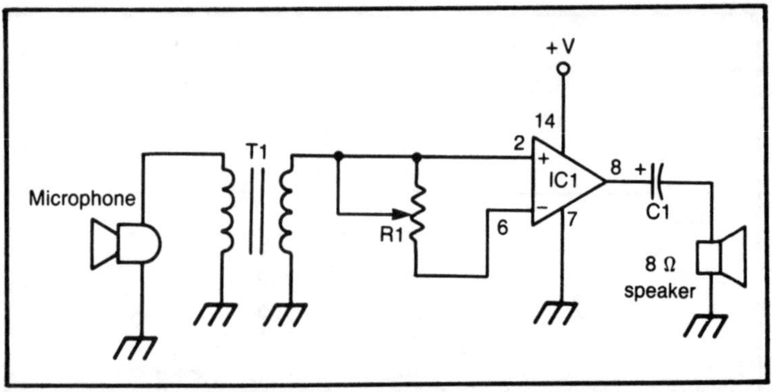

Fig. 9-6. A practical audio amplifier circuit using the LM380 (Project #27).

amplifier IC. The output decoupling capacitor is needed if a single-polarity power supply is used, as shown in the diagram.

The LM380 is often used in inexpensive tape recorders and phonographs. Figure 9-7 shows a simple phonograph amplifier circuit. The input in this case is a ceramic cartridge. The parts list is given in Table 9-2. Potentiometer R1 is a simple voltage-divider volume control. Potentiometer R3 is a basic tone control, making the high frequency roll-off characteristics of the circuit manually adjustable.

Most serious phonograph applications require frequency shaping to provide the standard RIAA equalization characteristic, discussed in Chapter 6. All commercially available records are equalized according to the RIAA standards. Obviously, if the complementary frequency response shaping isn't included in the playback, the reproduced sound won't be as good as it should be.

Figure 9-8 shows a LM380-based phonograph amplifier circuit with full RIAA equalization. The parts list is given in Table 9-3.

The mid-band gain can be defined with this formula:

$$G = (R1 + 150000) / 150000$$

Table 9-1. Parts List for Project #27.

Component Number	Description
IC1	LM380 audio amplifier IC
T1	Transformer—500 ohm:200 K
C1	500 μF 25 V electrolytic capacitor
R1	1 Meg potentiometer

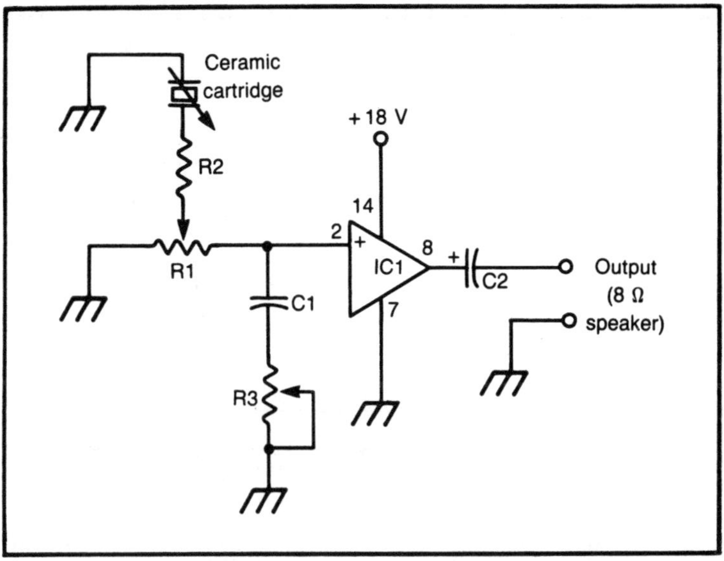

Fig. 9-7. The LM380 is ideal for use in inexpensive ceramic cartridge phonographs (Project #28).

The constant (150000) represents the internal resistance presented within the LM380 itself. R1 is the value of the external resistor.

The corner frequency is determined by resistor R1 and capacitor C1, according to this formula:

$$Fc = \frac{1}{2\pi C_1 R_1}$$

Table 9-2. Parts List for Project #28.

Component Number	Description
IC1	LM380 audio amplifier IC
C1	0.047 μF capacitor
C2	500 μF electrolytic capacitor
R1	25 K potentiometer
R2	75 K resistor
R3	10 K potentiometer

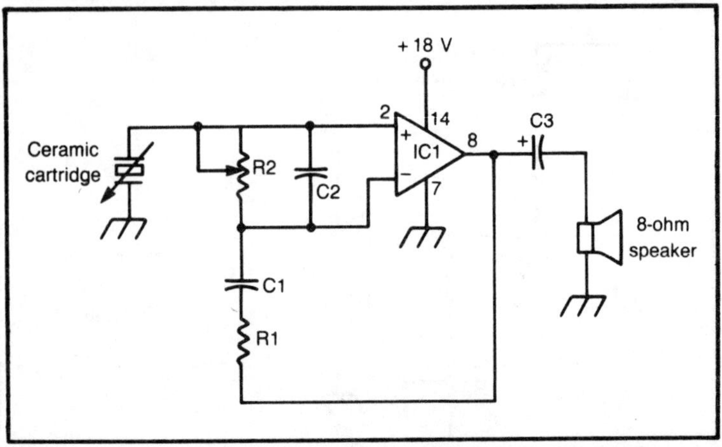

Fig. 9-8. This phonograph amplifier includes RIAA equalization (Project #29).

In designing such a circuit, it is usually best to select R1 for the desired gain first, then rearrange the corner frequency equation to solve for C1.

A pair of LM380 amplifiers can be put into a bridge configuration, as illustrated in Fig. 9-9. This is done to achieve more output power than could be obtained from a single amplifier. This circuit provides twice the voltage gain across the load for a given supply voltage, which increases the power handling capability by a factor of four over a single LM380.

When using the bridge configuration, caution is called for. The heat dissipation capabilities of the IC package can limit the maximum output power below the theoretical quadruple level.

Table 9-3. Parts List for Project #29.

Component Number	Description
IC1	LM380 audio amplifier IC
C1	220 pF capacitor
C2	0.0022 μF capacitor
C3	500 μF electrolytic capacitor
R1	1.5 Meg resistor
R2	2 Meg potentiometer

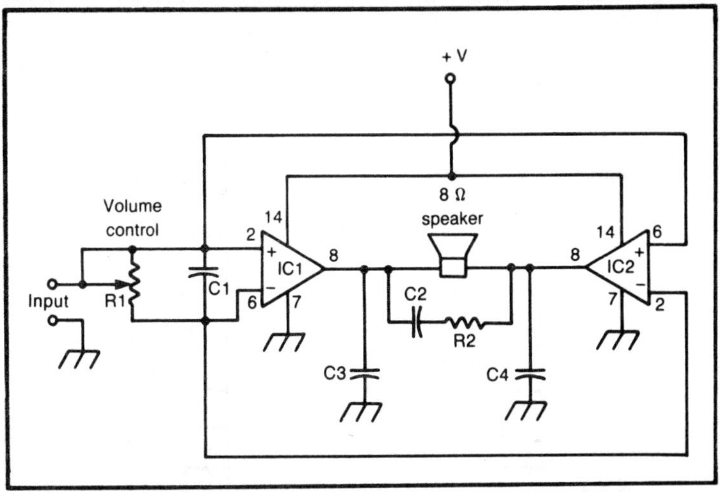

Fig. 9-9. A pair of LM380s in a bridge configuration can produce greater output power (Project #30).

A typical parts list for this bridged-amplifier circuit is given in Table 9-4.

This has barely scratched the surface of LM380-based amplifier circuits. You can see now why this chip is such a best-seller.

LM377/378/379 POWER AMPLIFIERS

Closely related to the LM380 audio amplifier IC is the LM377, LM378, and LM379 series of power amplifier ICs. The LM377 is a dual 2 watt amplifier. Its pin-out diagram is shown in Fig. 9-10. The LM378 is a dual 4 watt amplifier. Its pin-out is the same as for the LM377. Finally, the LM379 is a dual 6 watt amplifier, and its pin-out is shown in Fig. 9-11. Each of these ICs contains two

Table 9-4. Parts List for Project #30.

Component Number	Description
IC1, IC2	LM380 audio amplifier IC
C1	50 pF capacitor
C2, C3, C4	0.1 μF capacitor
R1	2 Meg potentiometer

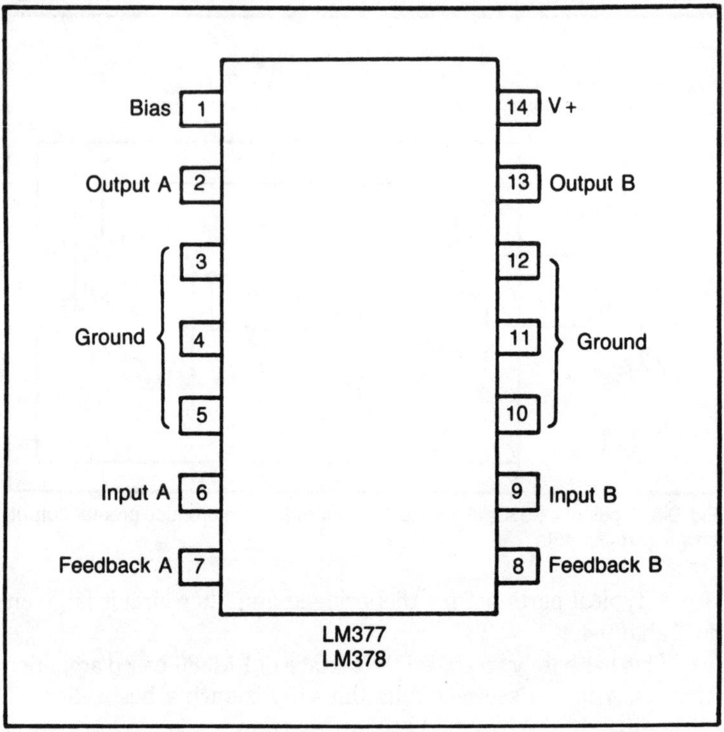

Fig. 9-10. The LM377 dual 2-watt amplifier IC and the LM378 dual 4-watt amplifier IC have identical pinouts.

complete amplifier stages, and they are ideally suited for stereophonic applications. According to the specification sheet, the separation between channels is a healthy 75 dB.

All of these chips are designed to drive 8 or 16 ohm loads directly. Since most modern loudspeakers present an 8 or 16 ohm impedance, no output transformers are normally required. In fact, very few external components are needed for most applications using one of these ICs.

Other features of the LM377/378/379 amplifier ICs include a typical open-loop gain of 90 dB, internal frequency compensation, 15 to 20 MHz gain-bandwidth product, and a fast turn on/turn off, without annoying (and potentially speaker-damaging) pops. The outputs of these chips are fully protected. Both output current limiting and thermal shut-down protection are provided.

Like the LM380, these amplifiers have both inverting and non-inverting inputs. The basic non-inverting circuit for these ICs is shown in Fig. 9-12. A typical parts list is given in Table 9-5.

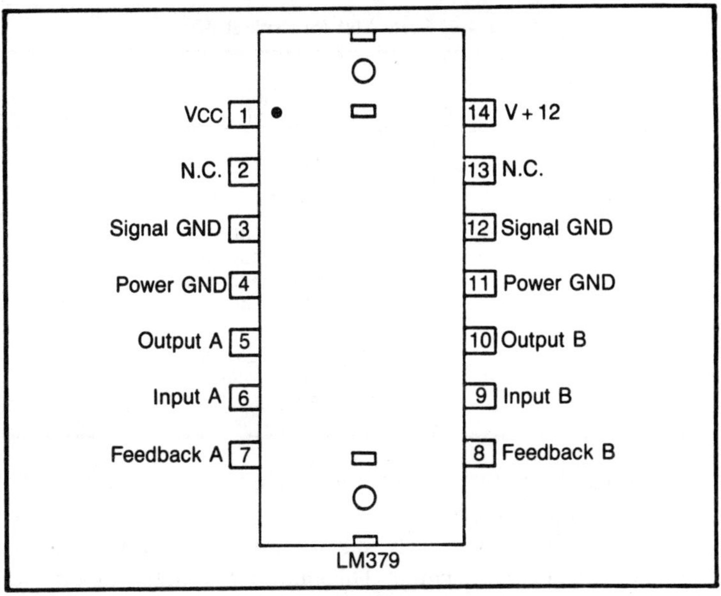

Fig. 9-11. The LM379 is a dual 6-watt amplifier IC.

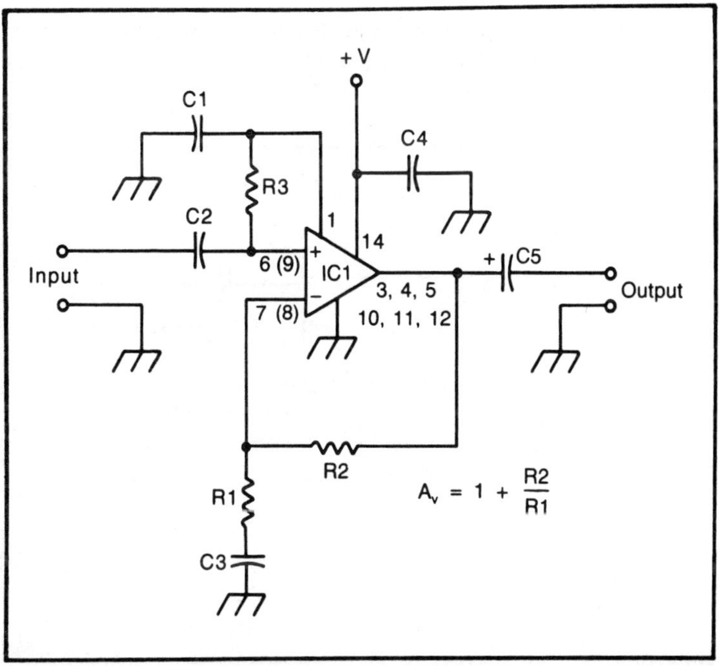

Fig. 9-12. There is no phase shift when the amplifier is used in the non-inverting configuration (Project #31).

Table 9-5. Parts List for Project #31.

Component Number	Description
IC1	LM377 preamplifier
C1, C3	0.47 μF capacitor
C2	0.01 μF capacitor
C4	0.1 μF capacitor
C5	500 μF electrolytic capacitor
R1, R3	22 K resistor
R2	1 Meg resistor

The basic inverting circuit is illustrated in Fig. 9-13, with the parts list given in Table 9-6.

As these diagrams clearly indicate, the inverting version has a smaller external parts count. This implies that inverting circuits tend to be less expensive when driven by a low impedance signal source.

Figure 9-14 shows how one of these ICs can be used to build a stereo amplifier in the non-inverting mode. The parts list for this project is given in Table 9-7.

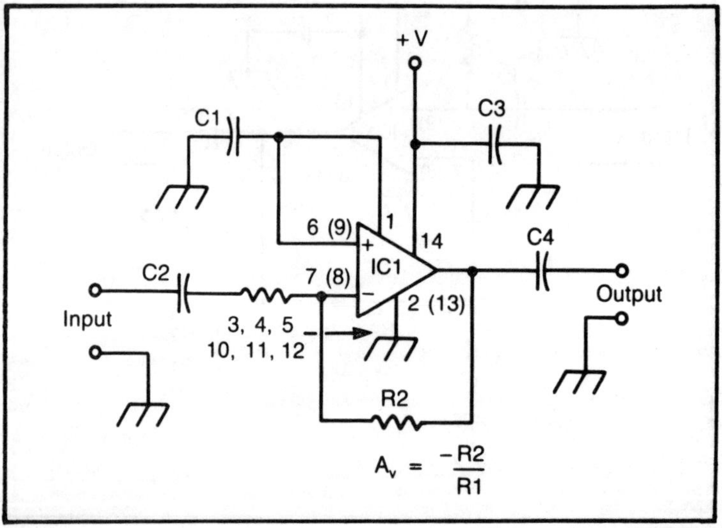

Fig. 9-13. In the inverting configuration, the signal is phase-shifted 180 degrees (Project #32).

Table 9-6. Parts List for Project #32.

Component Number	Description
IC1	LM377 preamplifier
C1	0.47 μF capacitor
C2	0.01 μF capacitor
C3	0.1 μF capacitor
C4	500 μF electrolytic capacitor
R1	select for desired gain
R2	select for desired gain

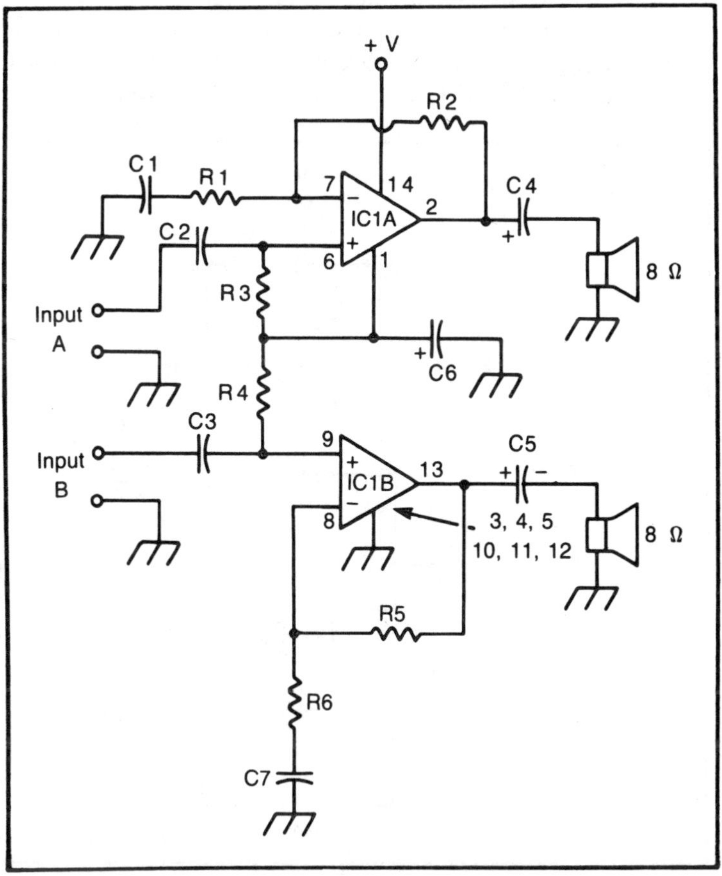

Fig. 9-14. A stereophonic non-inverting amplifier Project #33.

Table 9-7. Parts List for Project #33.

Component Number	Description
IC1	LM377 dual preamplifier IC
C1, C7	0.47 μF capacitor
C2, C3	0.01 μF capacitor
C4, C5	500 μF electrolytic capacitor
C6	330 μF electrolytic capacitor
R1, R6	22 K resistor
R2, R3, R4, R5	1 Meg resistor

An inverting stereo amplifier is illustrated in Fig. 9-15. The parts list is given in Table 9-8. The feedback resistors (rf) should not be any larger than about 1 Meg, or severe distortion may result.

The LM377/378/379 series of amplifier ICs places very little

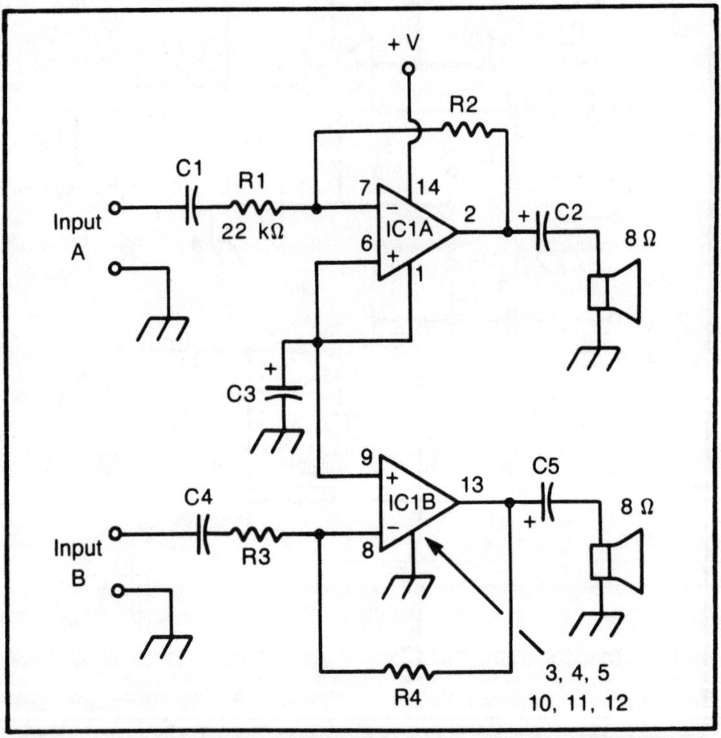

Fig. 9-15. A stereophonic inverting amplifier circuit Project #34.

Table 9-8. Parts List for Project #34.

Component Number	Description
IC1	LM377 dual preamplifier IC
C1, C4	0.47 μF capacitor
C2, C5	500 μF electrolytic capacitor
C3	330 μF electrolytic capacitor
R1, R3	22 K resistor
R2, R4	1 Meg resistor

loading on signal sources. The input impedance is approximately 3 Meg. Adding output transistors, as illustrated in Fig. 9-16 can significantly increase the power handling capability of these amplifiers. The LM378 normally puts out about 4 watts. In the circuit shown in Fig. 9-16, it produces 10 to 12 watts. Not only is this circuit quite simple for the power gained, it also exhibits even less crossover distortion than the LM378 alone.

THE μA783 AUDIO AMPLIFIER IC

It is really quite remarkable how much power today's IC designers are able to cram into a tiny chip of silicon. The μA783

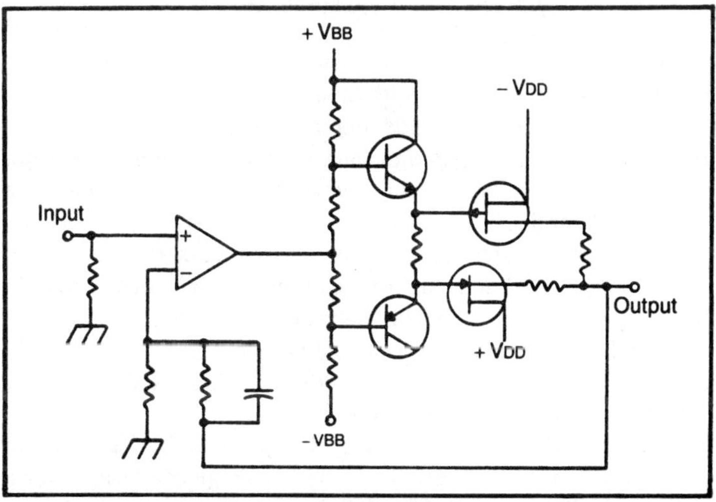

Fig. 9-16. The output power of an amplifier IC can be increased by adding external output transistors.

is a full 9 watt amplifier contained in a small 12 pin package that is less than an inch long and a quarter inch wide. In most applications for this IC, the largest thing in this circuit is the heat sink to prevent the IC from overheating.

The μA783 audio amplifier IC is designed to drive 8 or 16 ohm speakers directly. It is truly a high fidelity device with less than 0.3 percent of harmonic distortion.

THE LM381 LOW-NOISE DUAL PREAMPLIFIER

As discussed in Chapter 6, the signal from a magnetic tape head, or a phonograph cartridge (especially a magnetic type) is extremely small. The noise generated by the amplifier's circuitry itself could be even greater than the input signal. Obviously, too much amplifier noise could completely overpower the desired signal, rendering the entire system utterly useless.

Low-level input signals usually call for some sort of preamplifier. A preamplifier is nothing more than a low-power amplifier stage specifically designed for the best possible signal to noise (S/N) ratio. The preamplifier boosts the weak input signal up to a sufficient amplitude level at which it can be handled by the power amplifier.

A number of preamplifiers in IC form are available from various manufacturers. A fairly typical example is the LM381. This device contains two (stereo) high fidelity low-noise preamplifier stages in a single 14 pin package. The pin-out diagram for this IC is shown in Fig. 9-17. The LM381 is designed to generate a minimum of noise. A typical value for the internal noise generated by this device is 0.5 μV (0.0000005 volt) rms.

Two completely independent preamplifier stages are contained within the LM381. The two channels are very well isolated from each other. The channel separation specification for this chip is a respectable 60 dB.

An internal power supply decoupler/regulator results in a full 120 dB supply rejection. Any ripple in the supply voltage has little or no noticeable effect on the operation of the circuit.

The LM381 can also be run on a wide range of power supply voltages (9 to 40 volts). Like most modern amplifier ICs, the LM381 features output short circuit protection.

The LM381 provides a fairly hefty amount of gain for a preamplifier—up to 112 dB. The output voltage can swing over a wide range. The peak-to-peak voltage of the output can swing up to two volts below the supply voltage. For example, if the circuit

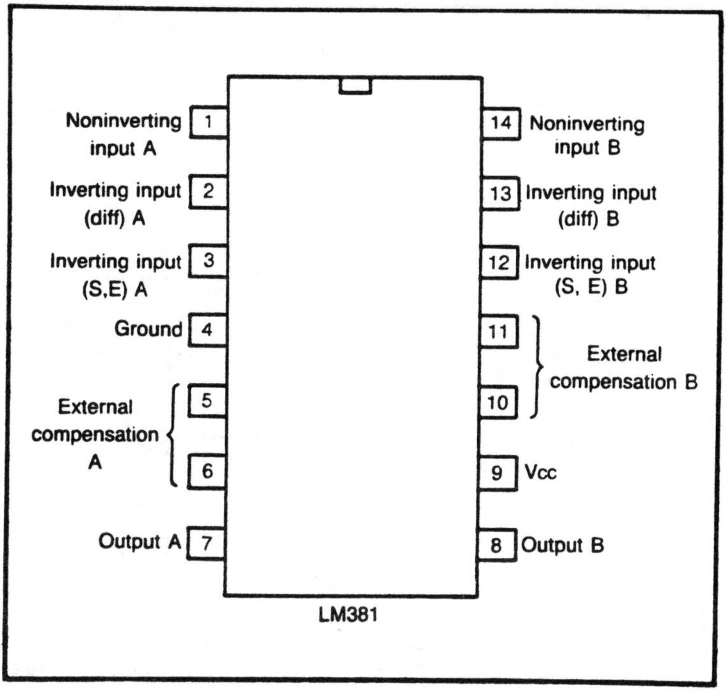

Fig. 9-17. The LM381 contains two high-fidelity low-noise preamplifier stages in a single 14-pin package.

is powered by a 12 volt supply, the output voltage can be as high as 10 volts peak-to-peak.

It should be apparent that this chip was designed primarily for high fidelity audio applications, but it is also quite suitable for wide-band instrumentation applications. The small signal bandwidth is 15 MHz, which carries the LM381's capabilities far past the upper limit of the audio range. To preserve stability in high-frequency applications, this IC includes internal frequency compensation.

A fairly typical application circuit for the LM381 preamplifier IC is shown in Fig. 9-18. This is a preamplifier for a magnetic tape head. This circuit is designed for more or less flat frequency response. The mid-band gain is determined by the resistor values, which is why these component values are not included in the parts list of Table 9-9.

The mid-band gain formula is:

$$Av = \frac{R_1R_2}{R_2}$$

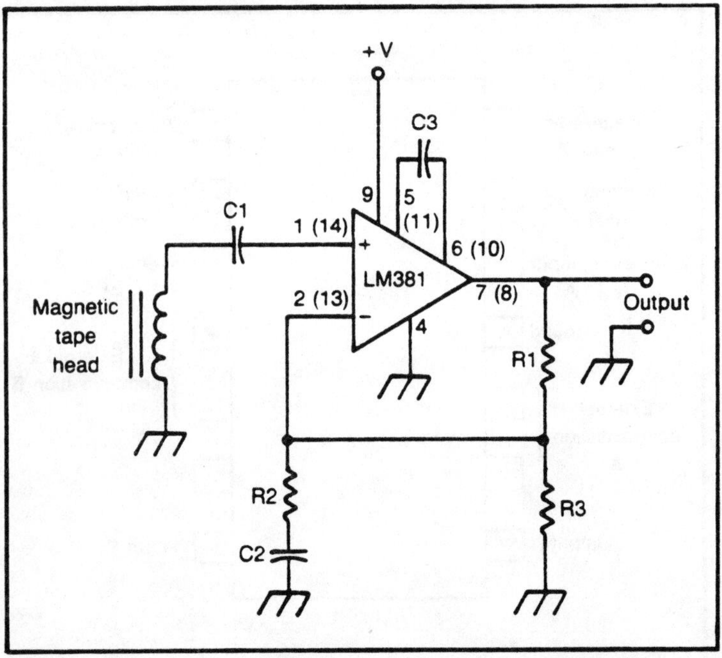

Fig. 9-18. The LM381 is a fine choice for a preamplifier in a tape recorder Project #35.

The low frequency 3 dB corner frequency is defined as:

$$fc = \frac{1}{2\pi\ C_2 R_2}$$

Table 9-9. Parts List for Project #35.

Component Number	Description
IC1	LM381 preamplifier IC
C1, C2	0.1 μF capacitor
R1	100 K resistor
R2	10 K resistor
R3	22 K resistor

Note that this formula is valid only if the capacitive reactance of C2 (Xc2) is equal to the resistance of R2.

The circuit of Fig. 9-18 did not incorporate any frequency compensation. Most modern tape recorders and players use standard NAB equalization, as discussed in Chapter 6. A modified version of this circuit is shown in Fig. 9-19. Here the LM381 is used in a NAB-compensated magnetic tape playback preamplifier.

Most good quality phonograph and tape player amplifiers include tone (bass and treble) controls to allow the listener to roughly customize the sound to his personal taste and/or the specific acoustic requirements of the listening environment. Generally, because of the inherent insertion loss of the tone control, it is usually located between two preamplifier stages. Thanks to the high gain capabilities and wide output range of the LM381, only a single preamplifier stage is normally required, even with completely passive tone control networks. Figure 9-20 shows a simple bass control network that is suitable for use with the LM381. Figure 9-21 shows a similar treble control network.

THE ICL8063 POWER TRANSISTOR DRIVER/AMPLIFIER

Most IC amps are designed for relatively low power applications. In applications requiring a heftier power output, external output transistors are usually mandatory. The ICL8063

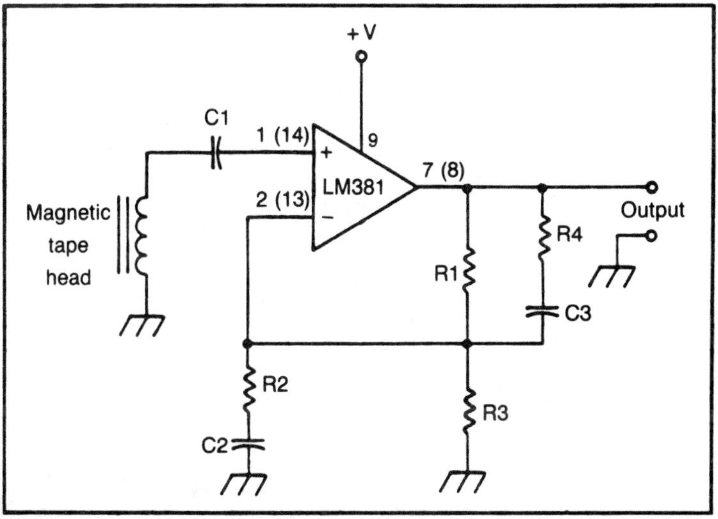

Fig. 9-19. This tape playback preamplifier includes NAB frequency compensation.

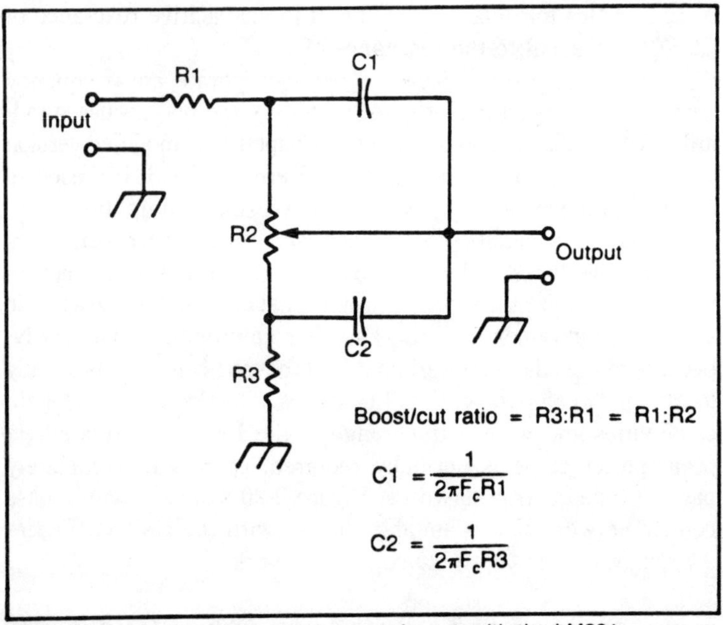

Fig. 9-20. A simple bass control network for use with the LM381.

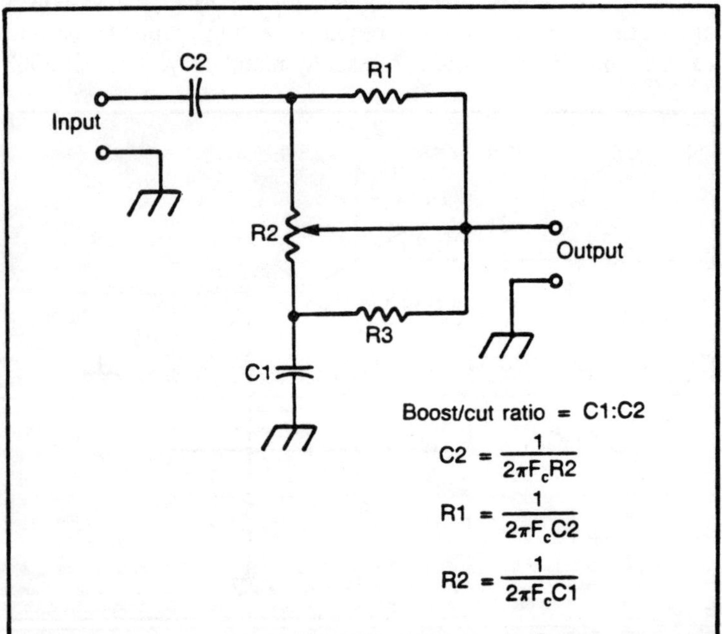

Fig. 9-21. A simple treble control network for use with the LM381.

is a monolithic amplifier specifically intended to drive power transistors for high-wattage outputs. The pin-out diagram for this device is shown in Fig. 9-22. This chip is intended to drive complementary-symmetry outputs in audio amplifiers. The ICL8063 can also be employed as a driver for servo and stepping motors, and rotary or linear actuators.

Basically, the ICL8063 takes signals in the ±11 volts range, and amplifies them up into the ±30 volts range at 100 mA. This increased level is suitable for driving power transistors.

The ICL8063 uses a dual-polarity 30 volt power supply. It includes on-chip ±13 volt regulators. This IC is designed to be fully compatible with most op amps, preamplifiers, companders, and similar IC devices.

Figure 9-23 shows a typical application for this device. The parts list is given in Table 9-10. Here the ICL8063 is being used in a 50 watt rms amplifier circuit, with an 8 ohm load (loudspeaker). Distortion is less than 0.1 percent for frequencies below about 100

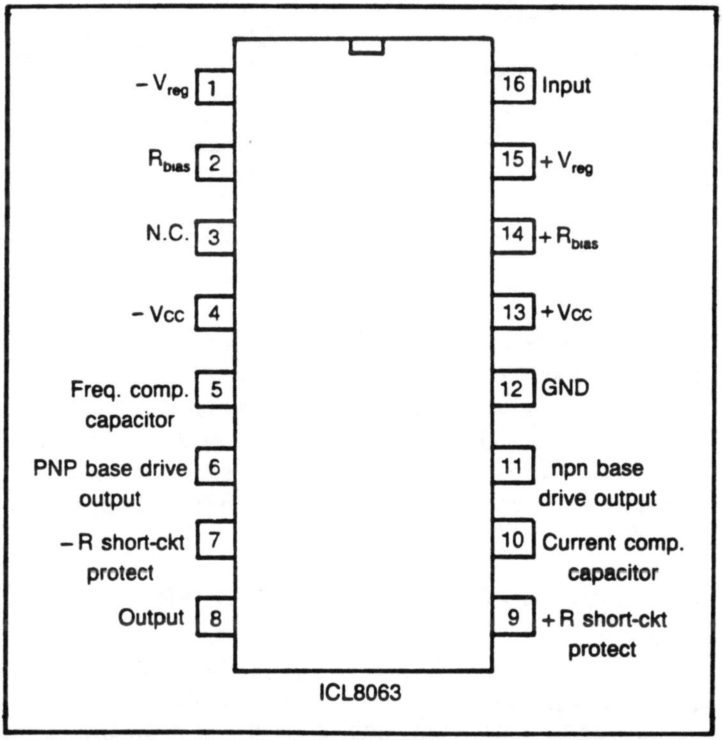

Fig. 9-22. The ICL8063 is designed to drive high power transistors for a high-wattage output.

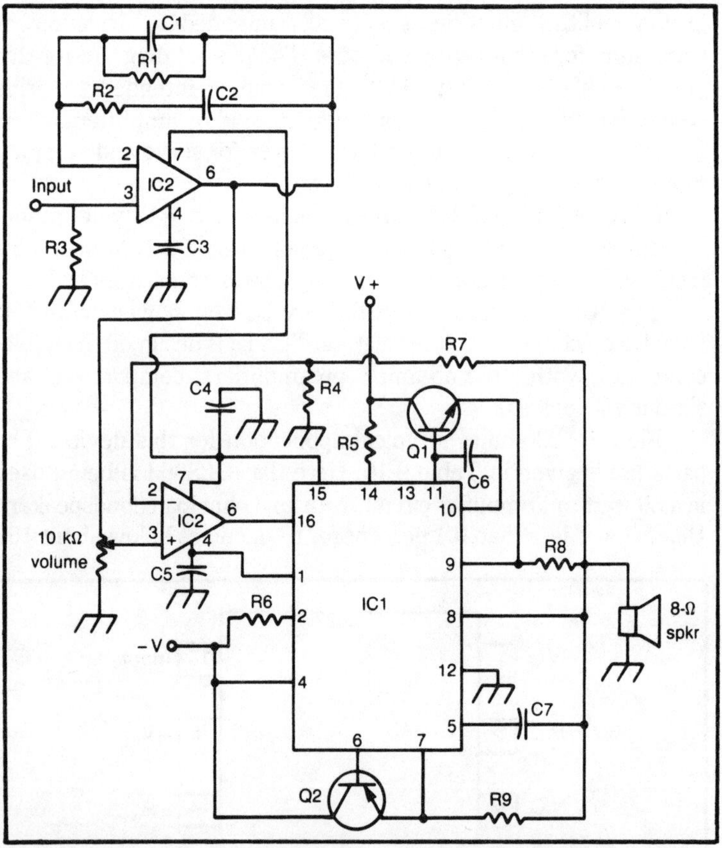

Fig. 9-23. This circuit puts out 50 watts into an 8 ohm load Project #36.

Hz. The distortion only goes up to about 1 percent at the high end of the audio spectrum (20 kHz).

The two 0.4 ohm resistors are included for current limiting. The 1000 pF capacitor helps maintain good stability down to unity gain. Of course, considering the relatively high currents passing through them, the transistors and the ICL8063 must be adequately heat sinked.

THE HA-2400 PROGRAMMABLE AMPLIFIER

A rather unusual amplifier IC is the HA-2400. This unique device is known variously as a programmable amplifier (PRAM) or a four-channel operational amplifier. This chip's pin-out diagram is shown in Fig. 9-24. Figure 9-25 illustrates the internal structure of this unit.

Basically, there are four op amp input stages within the HA-2400. Any one of those input stages can be selected via a digitally controlled electronic switch (pins 15 and 16). The selected input stage then drives a fifth op amp which serves as an output stage. Table 9-11 lists the various digital input combinations and their results. When the enable pin (14) is high, one of the four input stages is connected to the input of the output stage. Any standard op amp application (see Chapter 8) can be performed with the HA-2400, with the added advantage of programmability.

The output stage is internally wired as a unity gain voltage follower, so feedback components can be externally connected from the chip's output (pin 10) back to the appropriate input (pins 1 through 8). The HA-2400 functions just like a regular op amp, depending on which input stage is currently selected.

Each of the four input stages can be connected for different op amp applications, allowing the function to be digitally selectable.

Always remember that the unselected input stages can still constitute a load at the amplifier output and the signal input. The analog input terminals of an OFF channel draw the very same bias current as an ON channel. The input signal limitations must be

Table 9-10. Parts List for Project #36.

Component Number	Description
IC1	ICL8063 transistor driver/power amplifier IC
IC2, IC3	741 op amp IC
Q1	2N3055 transistor (or similar)
Q2	2N3791 transistor (or similar)
C1	3750 pF capacitor
C2	0.01 μF capacitor
C3, C5, C6	0.001 μF capacitor
C4, C7	1000 pF capacitor
R1	220 K resistor
R2	33 K resistor
R3	51 K resistor
R4	1 K resistor
R5, R6	1 Meg 5 percent 1/2 watt resistor
R7	4.7 K resistor
R8, R9	0.4 ohm 5 watt resistor

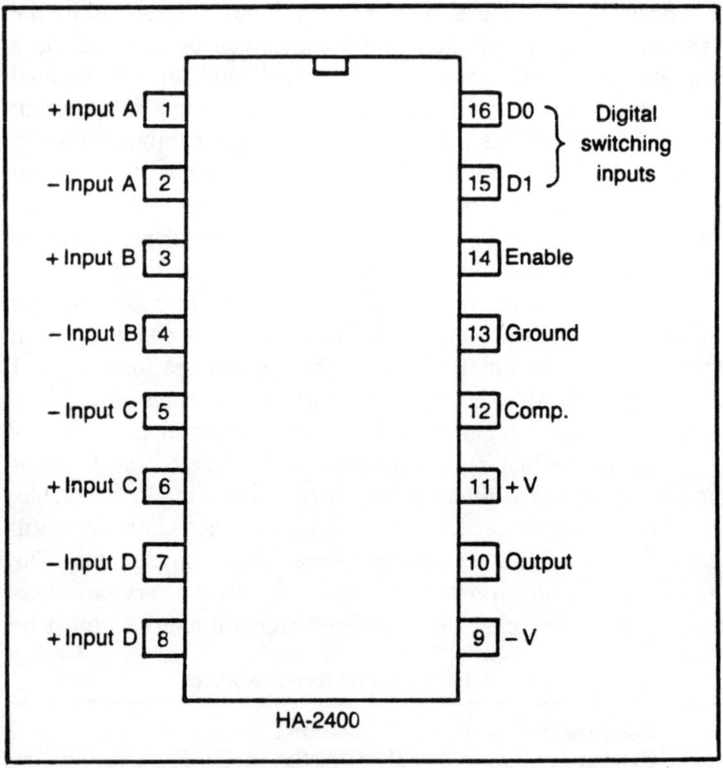

Fig. 9-24. The HA-2400 Programmable Amplifier IC is also called a four-channel op amp.

observed, even when the channel is off. When pin 14 (ENABLE) is made low (grounded), all four input channels are off. Under these circumstances, the output voltage at pin 10 tends to slowly drift towards the negative supply voltage rail (−V). If the specific application at hand requires a zero-volt output, wire one of the input channels as a non-inverting voltage follower with the non-inverting input grounded (input of 0 volts). Select this "dummy" channel instead of deactivating the ENABLE input of the chip.

It is generally not possible to wire the outputs of two or more HA-2400s together because the output impedance remains low, even when all inputs are disabled. One way to get around this problem is to use the compensation pin (12) as the output. The voltage at this terminal is about 0.7 volts higher than at the actual output pin (10), but the output impedance at pin 12 is very high. Consequently, two or more compensation outputs can be wired together without running into excessive loading problems.

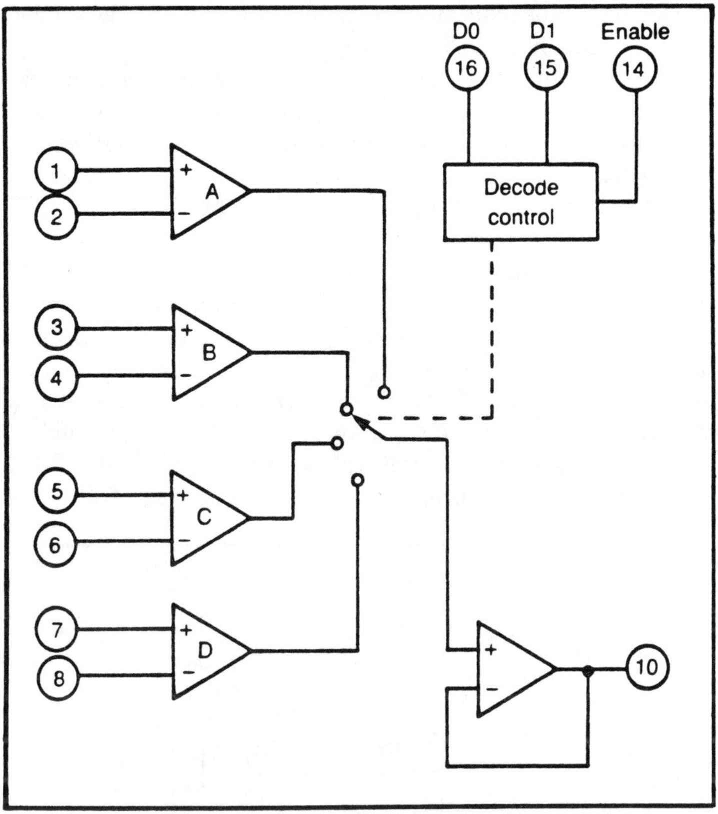

Fig. 9-25. The HA-2400 contains four digitally selectable input stages.

If the HA-2400 is being used with an overall closed loop gain of less than 10, external frequency compensation should be added to ensure closed-loop stability. This is usually done by connecting

Table 9-11. Different Digital Inputs Used to Select the Various Input Stages of the HA-2400 PRAM.

Digital Inputs			Channel			
D0	D1	Enable	A	B	C	D
X	X	0	Off	Off	Off	Off
0	0	1	Off	Off	On	Off
0	1	1	On	Off	Off	Off
1	0	1	Off	Off	Off	On
1	1	1	Off	On	Off	Off

a small capacitor (typically 2 to 15 pF) from pin 12 to ac ground. (The V+ supply is recommended by the manufacturer.)

Any unused digital inputs should be shorted to ground (for a permanent low state) or to +5.0 volts (for a permanent high state). If they are left floating, they behave as if held high, but may not be reliable.

The digital inputs of the HA-2240 are DTL and TTL compatible.

As you can undoubtably see, the HA-2240 is an extremely handy and economical device. The programmability offers a tremendous amount of versatility in circuit design. Even in applications where only a single channel is to be electronically switched on and off, it is often less expensive and more convenient to use a HA-2240 than a separate op amp and analog switch.

Figure 9-26 shows a typical application for the HA-2400 PRAM. The parts list for this project appears in Table 9-12. This circuit

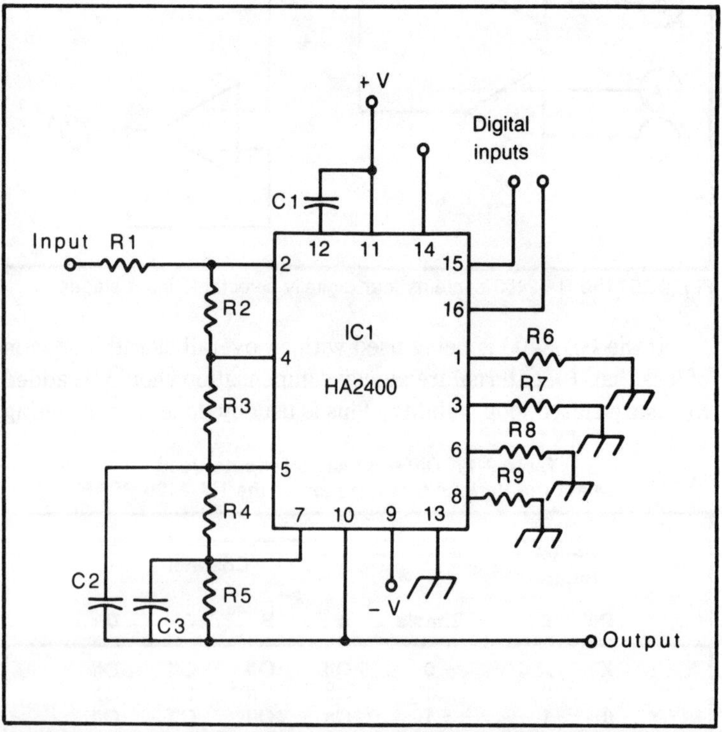

Fig. 9-26. This inverting amplifier circuit with programmable gain is a typical application for the HA-2400 PRAM Project #37.

Table 9-12. Parts List for Project #37.

Component Number	Description
IC1	HA2400 PRAM IC
C1	15 pF capacitor
C2, C3	5 pF capacitor
R1	44.6 K resistor *
R2	35.4 K resistor *
R3	53.4 K resistor *
R4	66.6 K resistor *
R5	200 K resistor *
R6	40 K resistor *
R7	64 K resistor *
R8	100 K resistor
R9	86.6 K resistor *

* Resistor values are exact values from calculations. May be approximated in non-critical applications.

is an inverting amplifier with programmable gain. Depending on the states of the digital inputs, the gain can be any of the following:

0 (OFF)
−1
−2
−4
−8

The same thing can also be done with one input resistor and one feedback resistor per channel, but by using a single feedback resistor, this circuit has a lower parts count; it uses only five resistors instead of eight.

Figure 9-27 shows the HA-2400 being used as a sine wave oscillator with a programmable output frequency. Notice how positive feedback is used to cause the amplifier to oscillate.

The circuit shown in Fig. 9-28 is a multi-function circuit. It can perform various addition and subtraction functions, depending on the controlling states of the digital inputs. The four possible functions that can appear at the output of this circuit are:

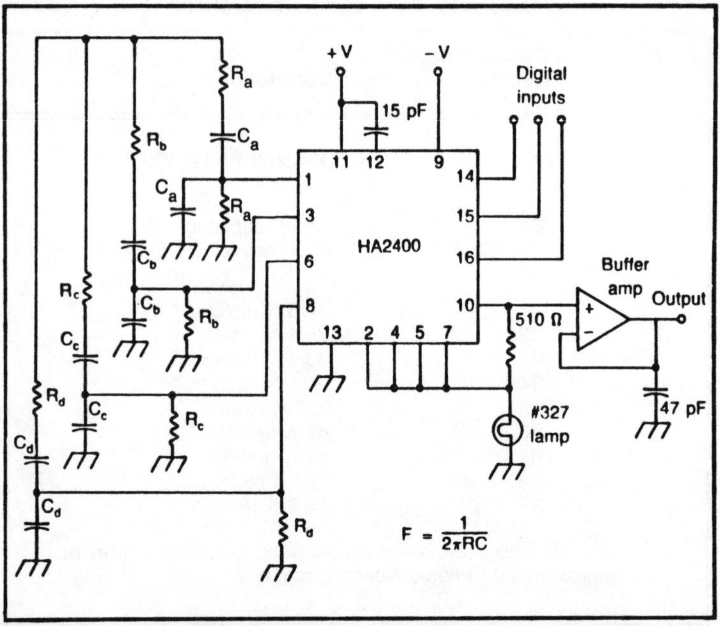

Fig. 9-27. The HA-2400 PRAM can also be used as a sine wave oscillator with a programmable output frequency.

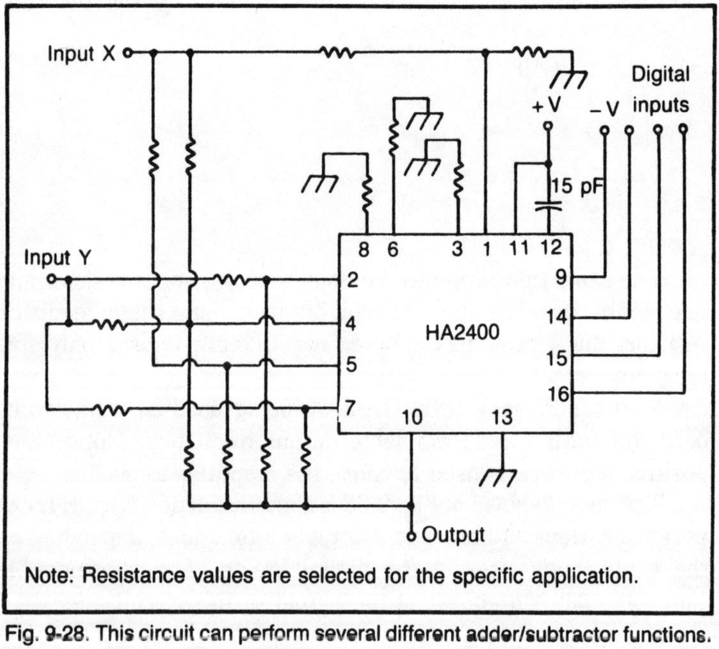

Fig. 9-28. This circuit can perform several different adder/subtractor functions.

- G1X
- G2Y
- (G3X + G4Y)
- G5X - G6Y

G stands for the appropriate gain, dependent on the resistance value used in the active channel.

These are just a few of the many novel applications for this fascinating IC.

RF AMPLIFIERS

Figure 9-29 shows the LM171 integrated rf/i-f amp IC. This chip can handle frequencies ranging all the way from dc up to 250 MHz. It can be emitter-coupled or cascaded according to the demands of the specific application.

All inputs and outputs are brought out to individual terminals on the IC, permitting great versatility in circuit design. Since the internal feedback of the LM171 is low, high stability is maintained.

Like most ICs designed for use in rf circuits, the LM171 is housed in a metal can, rather than the plastic body of many audio

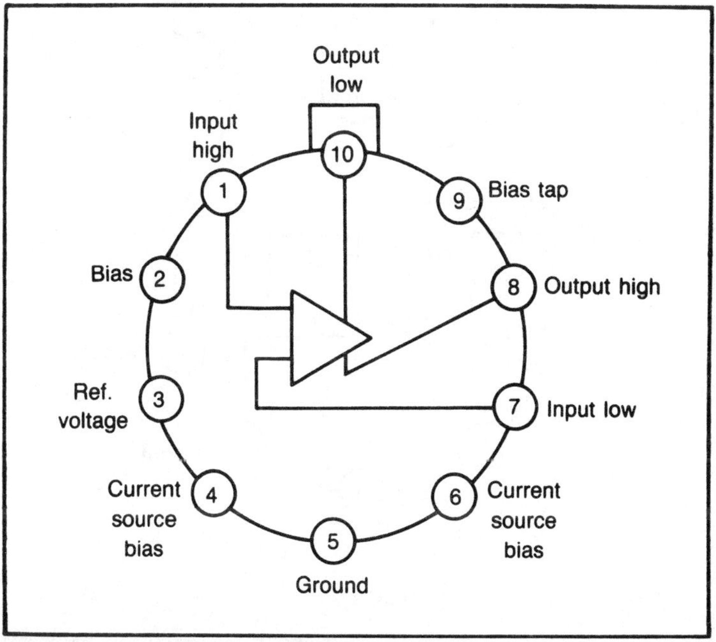

Fig. 9-29. The LM171 is a typical rf/i-f amplifier IC.

frequency ICs. The metallic case offers shielding against unwanted rf pick-up or transmission. Closely related to the LM171 the LM272 and the LM371 and a number of other rf amps are also available in IC form.

VIDEO AMPLIFIERS

Video amp ICs are also available. A fairly typical device of this type is the LM733, illustrated in Figs. 9-30 and 9-31. This IC is available either in a 10 pin metal can, or a 14 pin DIP housing.

The LM733's features include a 120 MHz band-width and 250 K input resistance to minimize loading the signal source. The gain can be selected from three fixed values:

10
100
400

No frequency compensation is normally needed.

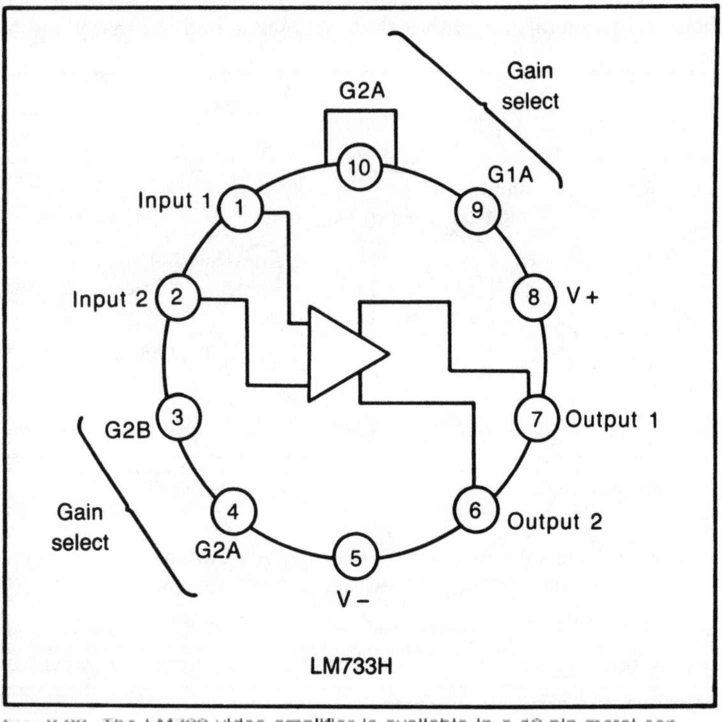

Fig. 9-30. The LM733 video amplifier is available in a 10-pin metal can.

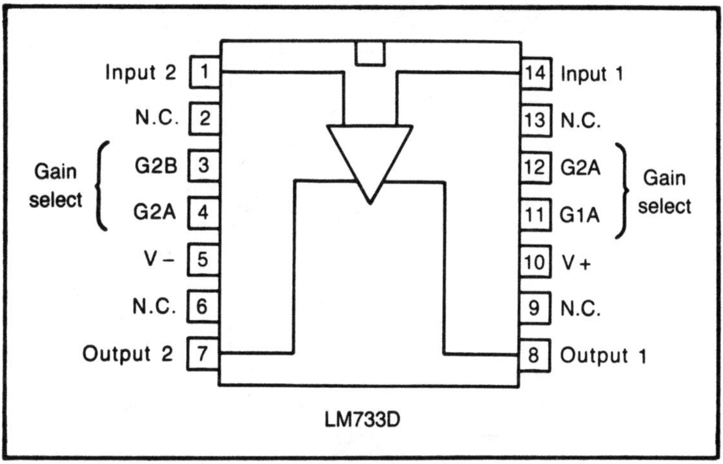

Fig. 9-31. The LM733 video amplifier is also available in a 14-pin DIP housing.

IN CONCLUSION

Hundreds of amplifier circuits are now available in IC form. We have obviously barely scratched the surface here; instead of being exhaustive, I have presented some typical and unique examples.

Anyone interested in circuit design should experiment with some of these IC amps. They can almost always be used in applications other than their designed function, if you use some ingenuity.

Chapter 10

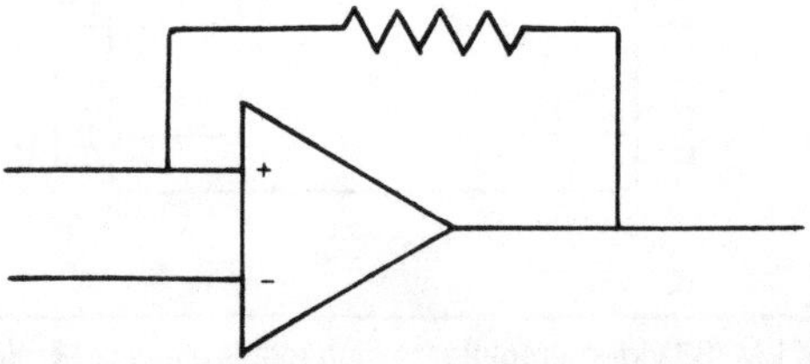

Voltage-Controlled Amplifiers

IN MOST AMPLIFIERS, THE GAIN IS EITHER FIXED OR MANUALLY variable, usually via a potentiometer. Generally, this is a perfectly logical approach and works just fine.

However, in a few specialized applications, there may be a need for variable gain, but for some reason manual control is not a practical or desirable option. Such applications include inaccessible or remote circuitry and certain precision and/or high speed operations. In such situations you need some means of controlling the gain of the amplifier by purely electrical means. A circuit of this type is called a Voltage-Controlled Amplifier, or VCA.

BASIC DESCRIPTION

A VCA has at least two separate inputs and a single output. One of the inputs is the usual signal input. The signal to be amplified is fed into the amplifier at this point. The other input accepts a control voltage. The signal fed to this input is usually a dc voltage, although ac control voltages are occasionally employed. The level of this control voltage determines the gain of the amplifier.

At the output you find a recreation of the original input signal amplified by an amount proportionate to the control voltage.

ELECTRONIC MUSIC

VCAs are used in many applications, but they are probably most obvious in electronic music synthesizers. There are a number of ways in which a voltage-controlled amplifier can be put to work in creating electronic music. A control voltage can adjust the output level faster and with far greater accuracy than any human being could possibly manage.

The four most common uses of VCAs in electronic music (which are also reflected in other applications) are:

- ☐ Gating
- ☐ Amplitude envelopes
- ☐ Tremolo
- ☐ Amplitude modulation

Gating is probably the simplest of these applications. A VCA can be used to temporarily block a continuous signal. Among other uses, this can help prevent annoying clicks that can be caused by quickly switching an audio signal source on or off. In a synthesizer, a gated VCA can also prevent any sound at the output when no key is being pressed on the keyboard.

The connections between the various modules in an electronic music synthesizer are called patches. In the earliest synthesizers, all such connections were physically made with patch cords, and the name has hung on to this day.

A typical gating patch is illustrated in Fig. 10-1. Notice that two outputs are taken off of the keyboard. One is a variable voltage that is used to control the VCO (Voltage-Controlled Oscillator). Each key produces a unique voltage. The other signal from the keyboard is simply a straight, unvarying voltage that appears at

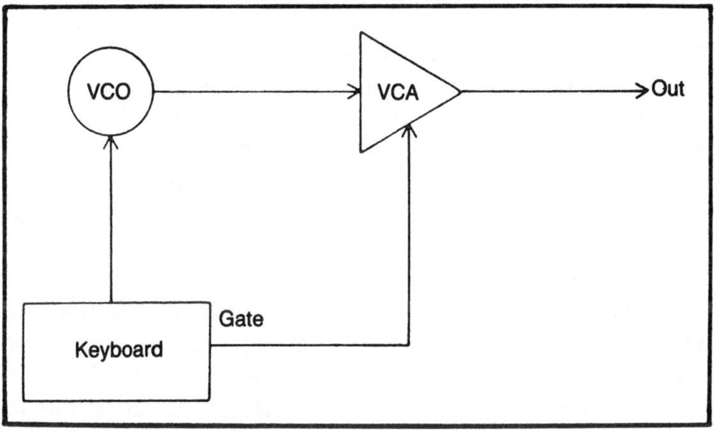

Fig. 10-1. A typical gating patch.

the gate output whenever one or more keys are being held down. If no keys are pressed, this output is at ground potential.

Most synthesizers also have a trigger output which produces a brief, fixed-duration pulse whenever a new key is initially pressed. The length of time the key is pressed does not affect the duration of the trigger pulse. But the gate output remains high as long as at least one key is being held down.

When no key is pressed in a simple patch the VCO continues to generate a signal based on the last key depressed. Actually, the frequency probably drifts downward when the key is released. Obviously that makes playing any kind of melodic passage exceedingly difficult, and rests (momentary silences) are impossible without manually switching the oscillator on and off.

But when no key is pressed, there is no gate signal, so the VCA's level is held to an output amplitude of 0, or silence. When a key is pressed, the gate signal turns up the volume on the VCA, and the tone is heard at the output as long as the key is held down.

In this arrangement, the VCA is either completely off (gain of 0), or at full amplitude (maximum gain). For all intents and purposes, it switches instantly between these two discrete levels.

While functional, this doesn't sound terribly natural, except on organ-like voices. Most natural sound sources take some finite amount of time to build up from the minimum (zero) level to the maximum level, and a finite time to die back down to the minimum level again. In many natural sounds, the amplitude goes up and down several times to varying degrees at specific rates that determine the characteristics of the sound.

These transition times between one amplitude level and another are extremely brief. Generally we don't even hear them as changes in amplitude at all, but these short transition times do have a very definite effect on the nature of the perceived sound. The changing amplitude pattern of a sound is called its amplitude envelope, and the audible effect is called timbre (pronounced tam-ber).

Timbre may be the most significant factor in determining what a sound is. The only major difference between the sound of a flute and the sound of a guitar is their amplitude envelopes.

An amplitude envelope always consists of at least two parts. The attack is the time it takes for the sound to build up from zero to its maximum level. The time it takes the sound to drop back down to zero is called the decay, or release. A few typical attack/release (AR) envelopes are illustrated in Fig. 10-2. A sound with the envelope shown in Fig. 10-2A has a very percussive effect. The

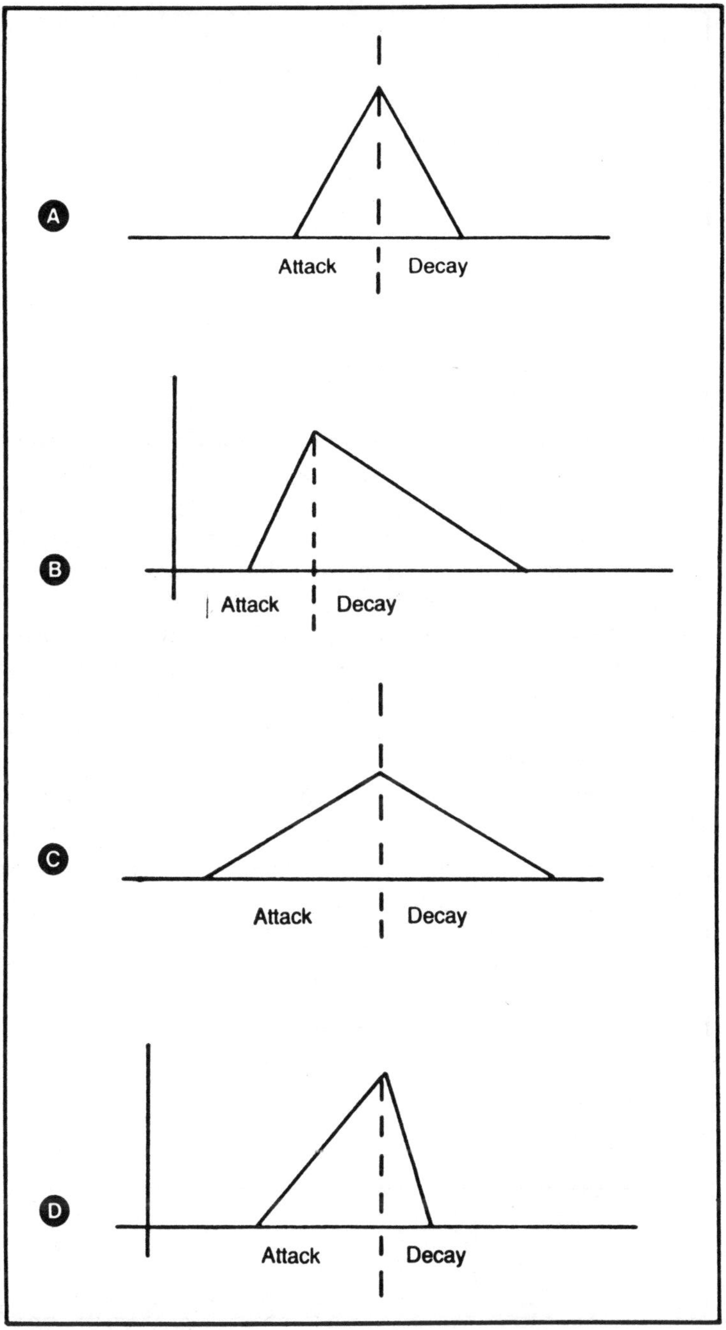

Fig. 10-2. Simple envelopes consist of an Attack and a Release.

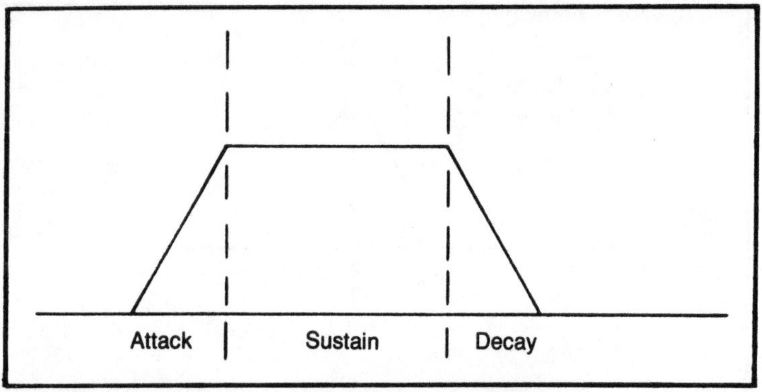

Fig. 10-3. A typical Attack-Sustain-Release envelope.

envelope in Fig. 10-2B has a rather short attack and a moderately long decay, like that of a guitar, or a piano. Figure 10-2C shows a long attack and a long decay, like a wind instrument of some kind. A long attack coupled with a short decay, as shown in Fig. 10-2D has a very unnatural, electronic-sounding effect. This type of amplitude envelope is not found in natural sounds.

Most sounds have more complex amplitude envelopes than the simple AR examples of Fig. 10-2. For one thing, many sound sources can hold the tone at the maximum level for some time. This is called sustain. A typical attack/sustain/release (ASR) envelope is illustrated in Fig. 10-3.

A still more complex amplitude envelope is shown in Fig. 10-4. First there is an attack, then an initial decay down to a specific

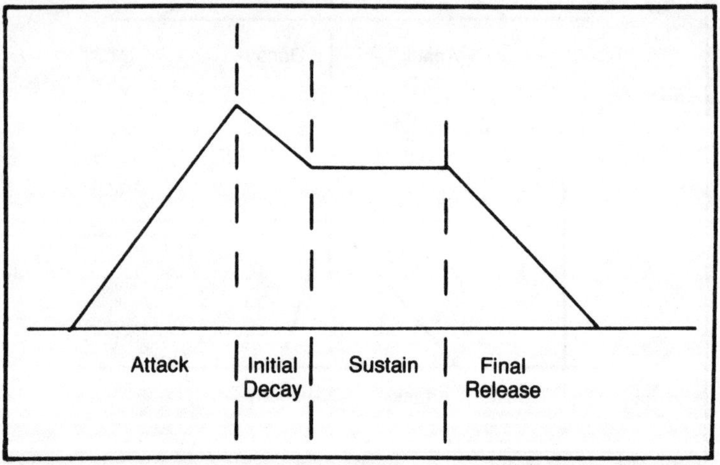

Fig. 10-4. Most electronic music synthesizers use ADSR envelopes.

sustain level somewhat below the maximum level. This sustain level is held until the key is no longer pressed, at which point the final release takes the amplitude back down to zero. This is called an Attack/Decay/Sustain/Release (ADSR) envelope. For convenience, the first declining portion of the envelope is called decay to distinguish it from the final declining section, which is usually referred to as the release.

The ADSR envelope is the most commonly used envelope in electronic music. Varying voltages that create these patterns are produced by special circuits called envelope generators, or occasionally, function generators.

Many natural sounds in the real world have extremely complex envelopes that may be difficult (though not necessarily impossible) to simulate electronically. An example of a complex amplitude envelope is shown in Fig. 10-5. Moreover, each harmonic in a complex tone may have a somewhat different amplitude. To synthesize such complex sounds, it is usually necessary to resort to a rather cumbersome procedure called additive synthesis.

Fortunately, the ear is not all that precise, especially when several sounds are heard simultaneously. In most cases, a standard simple AR, ASR, or ADSR envelope is usually sufficient. A trained ear can tell the difference, but the acoustic error is usually forgivable.

The basic envelope generator patch is shown in Fig. 10-6. The envelope generated can be controlled by either the keyboard's extended gate signal, or its brief trigger pulse. A gated envelope has a sustain portion to the envelope. A triggered envelope has no sustain.

If a low-frequency ac signal is applied to the control input of a VCA, an effect known as tremolo is the result. Tremolo is a

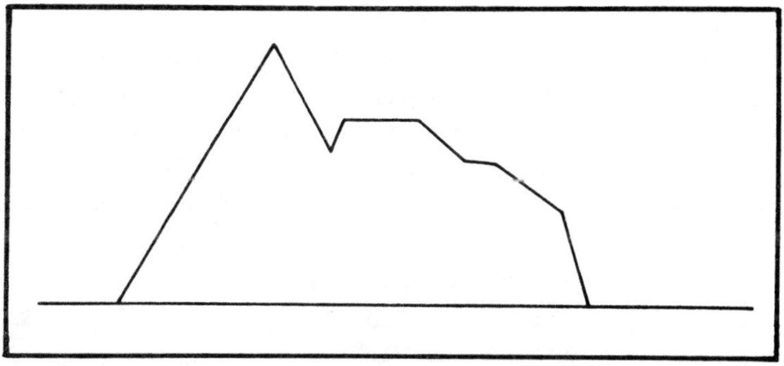

Fig. 10-5. Natural sounds often have very complex amplitude envelopes.

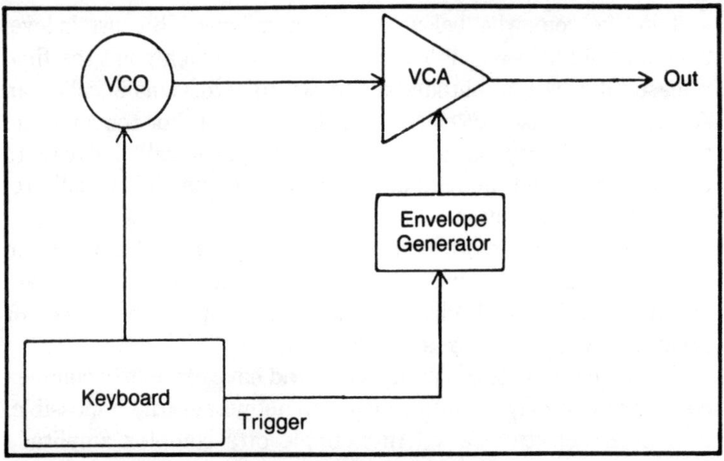

Fig. 10-6. The basic envelope generator patch.

regular audible fluctuation of the amplitude, producing a rather warbling quality. The sound is similar to but noticeably different from vibrato, which is a similar fluctuation of pitch.

For tremolo effects, the control voltage normally has a frequency between about 4 and 10 Hz. The amplitude of the control signal will determine the depth of the tremolo effect—that is, how much the amplitude of the output signal fluctuates.

If an ac control voltage is used with a higher frequency (in the 20 Hz to 20 kHz audible range), you get an interesting effect known as amplitude modulation. This effect is certainly not limited to electronic music. This is how AM radio signals are broadcast.

If the control voltage has an audible (or higher) frequency, it interacts with the signal at the input in a complex manner. Additional frequencies known as sidebands are created at the sums and differences of the control and input signals. The patch diagram for this effect is shown in Fig. 10-7.

For simplicity, let's assume that both signals are sine waves, with just a single frequency component each. Let's say that the input signal is a 2000 Hz sine wave, and the control signal is a 500 Hz sine wave. The output contains both of these frequencies, plus their sum and difference. In other words, the output signal is comprised of four frequency components:

500 Hz	Modulating frequency
1500 Hz	Difference
2000 Hz	Input signal
2500 Hz	Sum

Any pair of sine waves generate just two sidebands, regardless of the signal strength. When more complex waveforms are used for either or both of the VCA's inputs, a set of sum and difference frequencies is produced for each harmonic or other frequency component contained in either of the original signals.

As a second example, let's keep the control signal as a 500 Hz sine wave, but change the input signal to a 600 Hz square wave. To avoid too much complexity, assume a low-pass filter cuts off everything above 5000 Hz. As a result, the input signal is made up of the following frequency components:

600 Hz	Fundamental
1800 Hz	Third harmonic
3000 Hz	Fifth harmonic
4200 Hz	Seventh harmonic

The control signal sine wave modulates (creates a sum and difference frequency pair of sidebands) for each frequency component in the input signal. The sum and difference frequencies are summarized in Table 10-1.

All in all, the output signal in this example is made up of no fewer than thirteen frequency components:

100 Hz	Difference—fundamental
500 Hz	Modulating frequency
600 Hz	Fundamental
1100 Hz	Sum—fundamental
1300 Hz	Difference—third harmonic
1800 Hz	Third harmonic
2300 Hz	Sum—third harmonic

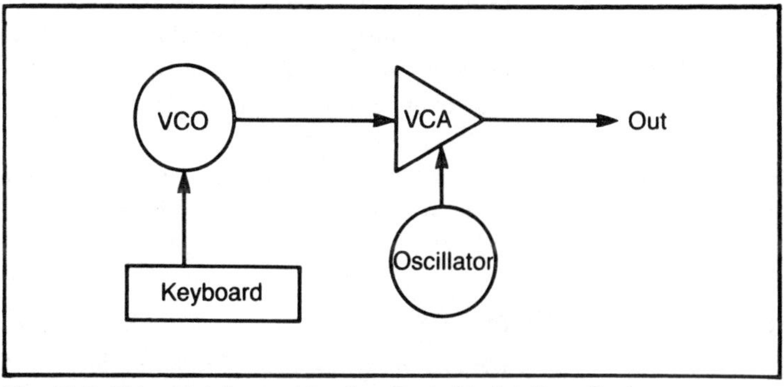

Fig. 10-7. This patch is used for Amplitude Modulation effects.

Table 10-1. A Summary of the Sum and Difference Frequencies Generated in the Amplitude Modulation Example Described in the Text.

Modulating frequency = 500 Hz

	Harmonic			
	Fundamental	**Third**	**Fifth**	**Seventh**
Program signal	600 Hz	1800 Hz	3000 Hz	4200 Hz
Difference (−500 Hz)	100 Hz	1300 Hz	2500 Hz	3700 Hz
Sum (+500 Hz)	1100 Hz	2300 Hz	3500 Hz	4700 Hz

2500 Hz Difference—fifth harmonic
3000 Hz Fifth harmonic
3500 Hz Sum—fifth harmonic
3700 Hz Difference—seventh harmonic
4200 Hz Seventh harmonic
4700 Hz Sum—seventh harmonic

If the modulating signal also consists of multiple frequency components, the situation becomes even more complex. For best results, at least one of the signals used in amplitude modulation should be a sine wave. More complex signals at both VCA inputs tend to create muddy sounding outputs with no definite sense of pitch. Of course, in dealing with electronic music, this may occasionally be exactly what you're looking for, in which case it's perfectly all right to break the rule.

Notice that unless the input signal and the control signal are harmonically related, the various overtones and undertones generated by the AM process are not harmonics of the perceived fundamental. In most cases, the perceived fundamental of an amplitude modulated signal is the same as the fundamental of the original input signal. However, if the modulating signal is very strong, its frequency may take precedence.

In dealing with amplitude modulation—especially in communications and radio work—the input signal is often called the program, and the modulating control signal is called the carrier.

SOME VCA CIRCUITS

Figure 10-8 shows what has to be one of the simplest VCA

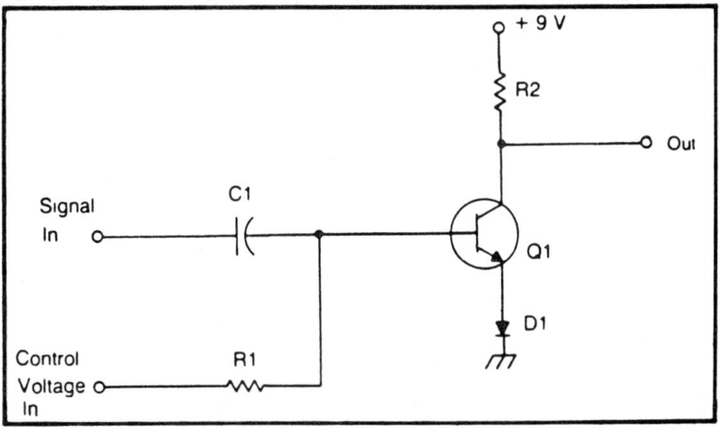

Fig. 10-8. A very simple VCA circuit Project #38.

circuits I have ever encountered. It is made up of just five simple components. The parts list is given in Table 10-2.

While any of a variety of different npn transistors may be used in this circuit, the diode must be a germanium type. (Silicon diodes are more common, but they do not work in this particular application.)

This circuit is really only suitable for experimentation. It has fairly poor low frequency response, so it is not recommended for use with frequencies below about 100 to 200 Hz. However, it does work pretty well with higher frequencies. The upper end of this circuit's frequency response extends well past the upper limit of audibility.

Two additional VCA circuits are illustrated in Fig. 10-9 and 10-10. Their parts lists are given in Tables 10-3 and 10-4, respectively. Both of these circuits are perfectly straightforward, and don't need any special comments.

Table 10-2. Parts List for Project #38.

Component Number	Description
R1	470 kΩ resistor
R2	22 kΩ resistor
C1	0.68 μF capacitor
D1	germanium diode (1N34A or similar)
Q1	NPN transistor (2N3564, GE-10, RCA SK 3019, Motorola HEP-54 Radio Shack RS-2011, or similar)

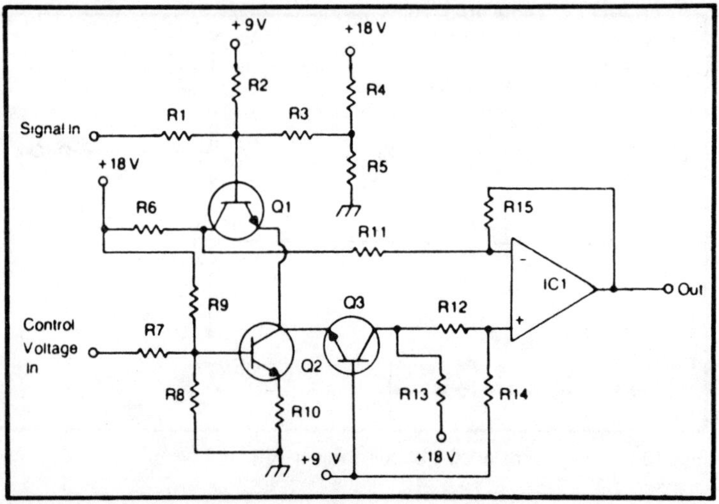

Fig. 10-9. An improved VCA circuit Project #39.

Not surprisingly, dedicated VCA ICs have appeared on the market. Curtis Electromusic Specialties, Inc. and Solid State Music are two manufacturers of VCA chips for use in electronic music synthesizers.

Figure 10-11 shows the pin-out for Solid State Music's SSM2010. Clearly, this is a fairly simple device to use. The pin functions are self-descriptive.

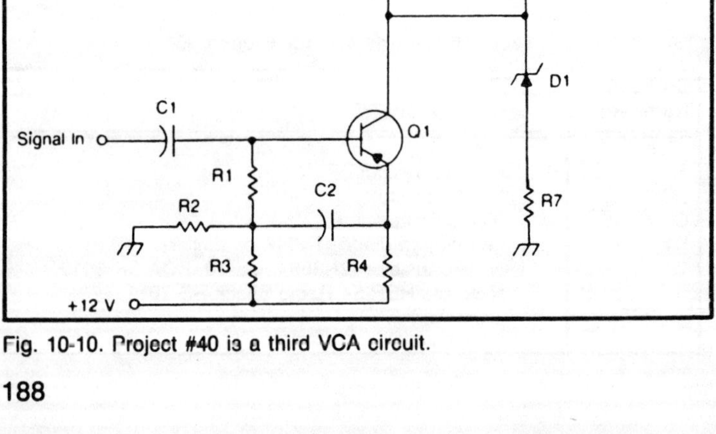

Fig. 10-10. Project #40 is a third VCA circuit.

Table 10-3. Parts List for Project #39.

Component Number	Description
R1	33 kΩ resistor
R2, R10	2.2 kΩ resistor
R3	10 kΩ resistor
R4	100 kΩ resistor
R5, R7	120 kΩ resistor
R6, R11, R12, R13	6.8 kΩ resistor
R8	22 kΩ resistor
R9	68 kΩ resistor
R14, R15	15 kΩ resistor
Q1, Q2, Q3	NPN transistor (2N3904, Radio Shack RS-2009, or similar)
IC1	low noise op amp IC

Table 10-4. Parts List for the Simple VCA Circuit of Fig. 10-10 Project #40.

Component Number	Description
R1, R5	1 kΩ resistor
R2	22 kΩ resistor
R3	4.7 kΩ resistor
R4	1.8 kΩ resistor
R6	120 Ω resistor
R7	56 Ω resistor
C1	0.1 μF capacitor
C2	1 μF capacitor
D1	6 V zener diode
Q1, Q2	PNP transistor (2N1384, GE-9, RCA SK3008, Motorola HEP-2, or similar)

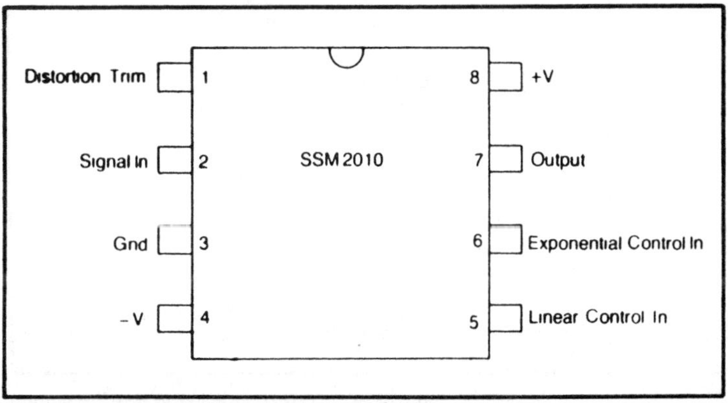

Fig. 10-11. The SSM2010 is a VCA in IC form.

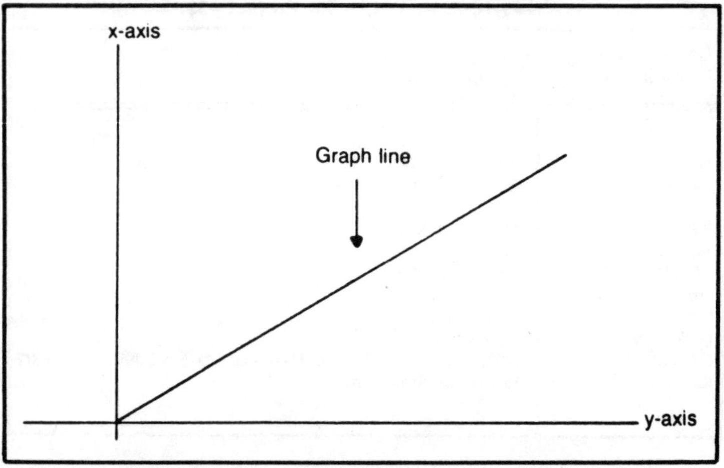

Fig. 10-12. Linear voltage control gives an obvious one-to-one relationship to the output amplitude.

The control voltage can provide either linear (see Fig. 10-12) or exponential (see Fig. 10-13) control over the amplitude. The human ear normally detects changes of volume in an exponential fashion, but linear control can also be useful for certain special effects.

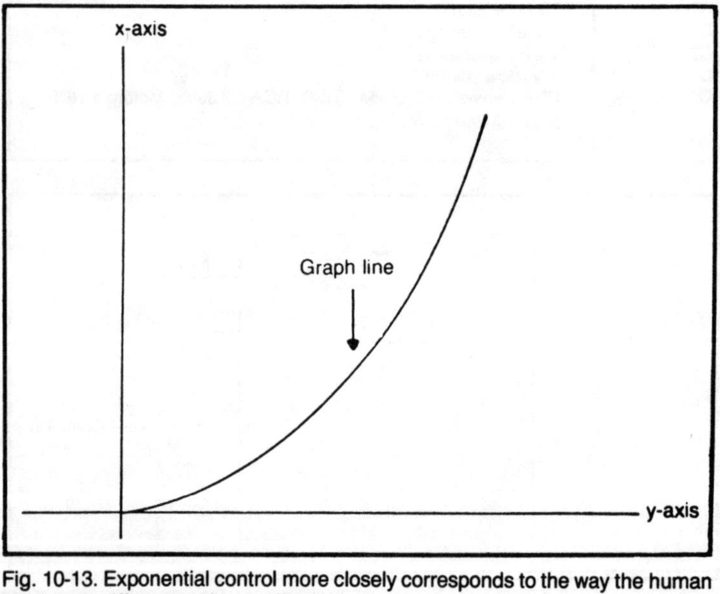

Fig. 10-13. Exponential control more closely corresponds to the way the human ear hears differences in amplitude.

REMOTE CONTROL

Another widespread, though slightly less obvious, application for voltage-controlled amplifiers is in remotely controlled devices. Some remote control functions are simply on/off type commands, but for other functions, a variable control may be required. For example, a remote control for a television set might include a volume adjustment. When this control is remotely activated, a signal is sent from the remote control to the set. The signal can be sent via a connecting cable, a beam of light, or a radio wave. In any case, the received signal at the set is in electrical form. A voltage proportionate to the received signal can be applied to the control input of a VCA to adjust the volume as instructed at the remote control location.

Index

Index

Edited by David Gauthier

Other Bestsellers From TAB

☐ **BASIC INTEGRATED CIRCUITS—Myles H. Marks**

With the step-by-step guidance provided by electronics design expert Myles Marks, anyone with basic electronics know-how can design and build modern IC devices for almost any application imaginable! In fact, this is the ideal sourcebook for every hobbyist or experimenter who wants to expand his digital electronics knowledge and begin exploring the fascinating new possibilities offered by today's IC technology. 432 pp., 319 illus.

Paper $16.95 Hard $26.95
Book No. 2609

☐ **EXPLORING LIGHT, RADIO AND SOUND ENERGY, with Projects—Calvin R. Graf**

Here's your chance to tune in to a whole new world of scientific phenomena . . . to build amplifiers and sensors that tap noises and sounds that you've never even dreamed existed. Using a simple soldering iron, low-cost, readily available parts, and the detailed schematics, and step-by-step directions provided by the author, you can make units that will keep you occupied for many intriguing hours. 210 pp., 123 illus.

Paper $10.95 Book No. 1758

☐ **101 SOUND, LIGHT AND POWER IC PROJECTS—Shoemaker**

At last! Here's an IC project guide that doesn't stop with how and why ICs function . . . it goes one step further to give you hands-on experience in the interfacing of integrated circuits to solve real-world problems. Projects include sound control circuits such as alarms and intercoms; light control projects from photoflash slave to a monitor/alarm; power control units and much more! 384 pp., 135 illus.

Paper $16.95 Book No. 2604

☐ **POWER SUPPLIES, SWITCHING REGULATORS, INVERTERS AND CONVERTERS—Gottlieb**

With such a large percentage of today's electronic equipment deriving their operating power from regulated supplies using inverters and converters, you owe it to yourself to stay in step with all the latest developments. This exceptional guide gives you all the technical data you'll need, from the basics of inverter, converter, and switching-type power supplies, to solid-state relays, logic-controlled supplies, cycloconverters, induction heating methods, linear regulators, and much more! All the data and hands-on guidance you need in one power-packed volume. 448 pp., 260 illus.

Paper $17.95 Book No. 1665

Other Bestsellers From TAB

☐ **TROUBLESHOOTING AND REPAIRING ELECTRONIC TEST EQUIPMENT—2nd Edition—Mannie Horowitz**

With this invaluable sourcebook as your guide you can troubleshoot and repair any test instruments. And if you've had a professional fix any of your test equipment lately . . . you have a pretty good idea how much money this book can save you! This comprehensive guide will bring you up-to-date on all the latest state-of-the-art troubleshooting and repair techniques. 448 pp., 218 illus.

Paper $17.95 Book No. 1663

☐ **UNDERSTANDING ELECTRONICS—2nd Edition—R. H. Warring**

Build a solar energy meter . . . a voltage regulator . . . a basic amplifier circuit . . . a single transistor radio with preamplification . . . a battery charger with stabilized output . . . and any of the dozens more exciting and practical projects included in this exceptionally complete guide to understanding and applying electronics in your home and workshop. Everything you need to get started is here in this exceptionally complete introduction to how and why electronics works. 210 pp., 19 illus.

Paper $9.95 Hard $15.95
Book No. 1553

*Prices subject to change without notice.

Look for these and other TAB books at your local bookstore.

TAB BOOKS Inc.
P.O. Box 40
Blue Ridge Summit, PA 17214

Send for FREE TAB catalog describing over 1200 current titles in print.

OR ORDER TOLL-FREE TODAY: **1-800-233-1128**